DATE DUE FOR RETURN		
23. JUN 98		

This book may be recalled
before the above date

the
RAINBOW
and the WORM

The Physics of Organisms

Mae-Wan Ho

Bioelectrodynamics Laboratory
The Open University
UK

the
RAINBOW
and the WORM

The Physics of Organisms

World Scientific
Singapore • New Jersey • London • Hong Kong

Published by

World Scientific Publishing Co. Pte. Ltd.
P O Box 128, Farrer Road, Singapore 9128
USA office: Suite 1B, 1060 Main Street, River Edge, NJ 07661
UK office: 73 Lynton Mead, Totteridge, London N20 8DH

100062008S

THE RAINBOW AND THE WORM
The Physics of Organisms

ISBN 981-02-1486-3
ISBN 981-02-1487-1 (pbk)

Printed in Singapore.

PREFACE

This book began 30 years ago when as a young undergraduate, I came across Szent Györgyi's idea that life is interposed between two energy levels of the electron. I was so smitten with the poetry in that idea that I spent the next 30 years searching for it. Naturally, I went at once into biochemistry. But I arrived too late. For the heady days of unity in diversity, the wine-and-roses days of discoveries in the universal energy transformation processes in living organisms had long been eclipsed by the new excitement over the molecular basis of heredity: the structure of the DNA double-helix and the cracking of the genetic code whereby the linear structure of DNA specifies the linear structure of proteins. So I became a geneticist; but found to my dismay that no one cared about unity anymore. Instead, almost everyone was obssessed with diversity and specificity: each biological function has a specific protein encoded by a gene 'selected by hundreds of milllions of years of evolutionary history' to do the job.

After some years getting acquainted with a group of enzymes involved in a number of hereditary neurological disorders, I became an evolutionist. Together with Peter Saunders, we set out to try to understand the relevance of thermodynamics to evolution, and other 'big' questions such as how life began. Sidney Fox's work on thermal proteins was the antithesis to the then dominant 'frozen accident' theory of prebiotic evolution, and impressed on me the nonrandomness, or I should say, non-arbitrary nature of life's physicochemical beginnings. I spent the next years working towards an alternative research programme to the neo-Darwinian theory of evolution by the natural selection of random mutations. This involved investigations on the relationship between development and evolution, the physicochemical and mathematical bases of biological form and organization, on 'rational taxonomy' - the classification of biological forms based on the developmental process generating the forms, as well as the wider social implications of evolutionary theory; all in the inspiring company of Peter Saunders, and later, also Brian Goodwin.

Yet something was still missing. I still did not understand how catching the electron between the excited and the ground state could make life go round. A chance meeting with Fritz Popp in 1985 changed all that. Although I understood then almost not a word of his lecture, I was convinced what he said was very significant: that coherence held the key to living organization. And so I plunged into quantum physics, with the help and encouragement of Fritz, who also invited me to work on light emission from living organisms in his laboratory. With immense patience and panache, he taught me quantum physics, taking me straight into the exotic world of quantum optics. Having got in at the deep end, I could do little else but swim to shore! Suddenly, everything begins to make sense.

When Brian Goodwin said to me: "Why don't you write a book on what is life?" I thought, why not? A month later, I put pen to paper, and the first draft was complete by the time another month had elapsed.

This book is patterned - roughly - after Schrödinger's *What is life?* (Cambridge University Press, 1944), and addresses the same question: Can living processes be explained in terms of physics and chemistry? His preliminary answer, in the 1940s, is that the inability of the physics and chemistry of his day to account for living processes is no reason for doubting that living processes could be so explained. For, on the one hand, there has not been an adequate experimental investigation of *living* processes, and on the other, physics and chemistry are both evolving disciplines, and some day, they *may* just succeed in accounting for life.

Indeed, physics and chemistry have developed a great deal since Schrödinger's time. Whole new disciplines became established: synergetics, nonequilibrium thermodynamics, quantum electrodynamics and quantum optics, to name but a few. There have already been several suggestions that the new physics and chemistry may be particularly relevant for our understanding of biological phenomena. I, for one, believe it is time to examine Schrödinger's question again, which provides a ready structure for this book.

I shall be ranging widely over the different physical disciplines already mentioned, starting from first principles; as well as the relevant physiology, biochemistry and molecular biology of cells and organisms. Those who expect details of molecular genetics and gene control mechanisms, however, will be disappointed. They should consult instead the numerous volumes of excellent texts that already exist. In *this* book, it is the bare-bones sketch that I shall be concerned with - the fundamental physical and chemical principles which are relevant for life.

At the risk of giving the story away, I shall say a few words on what this book is about. The first chapter outlines the properties of organisms that constitute 'being alive', and hence that which is in need of explanation. Chapters 2 to 5 show how a living system differs from a conventional thermodynamic machine. The organism's highly differentiated space-time structure is irreconcilable with the statistical nature of the laws of thermodynamics. And hence, the laws of thermodynamics cannot be applied to the living system without some kind of reformulation. The space-time structure of living organisms arises as the consequence of energy flow, and is strongly reminiscent of the non-equilibrium phase transitions that can take place in physicochemical systems far from thermodynamic equilibrium. Energy flow organizes and structures the system in such a way as to reinforce the energy flow. This organized space-time structure suggests that both quasi-equilibrium and non-equilibrium descriptions are applicable to living systems, depending on the characteristic times and volumes of the processes involved. Hence, an appropriate thermodynamics of the living system may be a *thermodynamics of organized complexity* involving, among other things, replacing the concept of *free* energy with that of *stored* energy.

Thermodynamics has its origin in describing the transformation of heat energy into mechanical work. As shown in Chapters 6, 7, 8 and 9, the predominant energy transductions in the living system are instead, electronic, electric and electromagnetic, as consistent with the primary energy source on which life depends as well as the electromagnetic nature of all molecular and intermolecular forces. Furthermore, given the organized, condensed state of the living system, it is predicted that the most general conditions of energy pumping would result in a phase transition to a dynamically coherent regime where the whole spectrum of molecular energies can be mobilized for structuring the system, and for all the vital processes which make the system alive. Chapters 8 and 9 summarize experimental evidence for coherence in living organisms, some of the findings pushing at the frontiers of quantum optics and quantum electrodynamics. The relevance of quantum theory to coherence in living systems is treated in more detail in Chapter 10.

The summary of the enquiry suggests that organisms are *coherent* space-time structures maintained macroscopically far from thermodynamic equilibrium by energy flow. This has profound implications on the nature of knowledge and knowledge acquisition, as well as on the nature of time as it relates to issues of determinism and freewill, which are dealt with in Chapters 11 and 12. There, I try to show how, by following to logical conclusions the development of western scientific ideas since the beginning of the present century, we come full circle to validating the participatory framework that is universal to all traditional indigenous knowledge systems the world over. This enables us to go some way towards restoring ourselves to an authentic reality of nonlinear, multidimensional space-time as experienced by the truly participatory consciousness, who is also the repository of free will and coherent action.

I have been asked whether it is necessary to do the science before getting to the point, which is the validation of participatory knowledge. Why not just develop the 'rhetoric'? That would have defeated one of the main purposes of this work, which is to show how science can reveal in a precise way the deeper wonders and mysteries of nature which are currently in danger of being totally obscured by the kind of superficial woolly misrepresentation that many people, especially the young, mistake to be 'new age philosophy'. Another motivation for writing this book is to show that science, when properly perceived, is far from being alienating and dehumanizing. Instead, it is consonant with our most intimate experience of nature. To me, science is surely *not* about laying down eternal 'laws of nature' to dictate what we can or cannot think. It is to initiate us fully into the poetry that is the soul of nature, the poetry that is ultimately always beyond what theories or words can say.

This book is the most exciting thing I have ever written. I would like to dedicate it to everyone young or youthful who has ever looked upon the living process in wonderment. I cannot express my gratitude enough to all those already mentioned, who have both inspired and encouraged me in realizing this work. Peter Saunders, as always, gives up much of his time to help me with the mathematics and the physics.

Very special thanks are due to my former colleagues, Oliver Penrose (Professor of Applied Mathematics and Theoretical Physics at the Open University, now at Herriot-Watt University, Edinburgh) and Kenneth Denbigh (Prof. of Physical Chemistry and Principal at Queen Elizabeth College, London) for reading an earlier version of the entire manuscript, commenting in substantial detail, correcting errors in my presentation, and stimulating me to new levels of understanding. To my present colleague, Nick Furbank (Professor of English at the Open University), I owe the title of this book and much more: he read and gave me helpful suggestions on the penultimate draft, and restored my faith that the book will reach a wider readership, including those without any scientific training who nevertheless love the subject as much as I do. Comments, suggestions and corrections are also gratefully acknowledged from Ronald Pethig, R.J.P. Williams, Clive Kilmister, Geoffrey Sewell, Wolfram Schommers, Francisco Musomeci, T.M. Wu, K.H. Li, R. Neurohr, Konrad Kortmulder, Lev Beloussov, Hansotto Reiber, Charles Jencks, Willis Harman, Lynn Trainor, Guiseppe Sermonti, Sheila Higham, Viven Thomas, and Katy Bentall. Last, but by no means the least, I would like to thank my brother, Li Poon and my son Adrian Ho for giving me their support as well as valuable comments and suggestions. The shortcomings which remain are all my own.

Milton Keynes,
July, 1993

CONTENTS

CHAPTER ONE

WHAT IS IT TO BE ALIVE?

The 'Big' Questions in Science

There are 'big' questions and 'small' questions in science. Most scientists in their work-a-day life confine themselves to asking small questions such as: Which gene is involved in a given hereditary defect? How will a certain organism react to such and such a stimulus? What are the characteristics of this or that compound? What is the effect of A on B? How will a given system behave under different perturbations? Yet, it is not a desire to solve particular puzzles that motivates the scientist, but rather the belief that in solving those little puzzles, a contribution will be made to larger questions on the nature of metabolic or physiological regulation, the generic properties of nonlinear dynamical systems, and so on. It is ultimately the big questions that arouse our passion - both as scientists and as ordinary human beings. They can inspire some of us as the most beautiful works of art that nature has created, whose meaning is to be sought as assiduously as one might the meaning of life itself.

For me, the big motivating question is Erwin Schrödinger's 'What is life?'[1] That it is also a question on the meaning of life is evident to Schrödinger, who closes his book with a chapter on philosophical implications for determinism and freewill. This is as it should be. I do not agree with those scientists for whom scientific knowledge has no meaning for life, and must be kept separate from real life in any event; perhaps an attitude symptomatic of the alienation that pervades our fragmented, industrial society. I will not dwell on that here, as it is not the main thesis of my book. Instead, I want to concentrate, for now, on Schrödinger's original question:

2

"How can the events *in space and time* which take place within the spatial boundary of a living organism be accounted for by physics and chemistry?"[2]

The same question has been posed in one form or another since the beginning of modern science. Is living matter basically the same as non-living only more complicated, or is it something more. In other words, are the laws of physics and chemistry necessary *and sufficient* to account for life, or are additional laws outside physics and chemistry required. Descartes is famous not only for separating mind from matter; he also placed living matter, alongside with non-living matter, firmly within the ken of the laws of physics; more specifically, of mechanical physics. Since then, generations of vitalists, including the embryologist Hans Driesch, the philosopher Henri Bergson, and the physiologist J.S. Haldane (the father of Marxist geneticist, J.B.S. Haldane), have found it necessary to react against the mechanical conception of life by positing with living organisms an additional *entelechy*, or *elan vital*, which is not within the laws of physics and chemistry[3].

The vitalists were right not to lose sight of the fundamental phenomenon of life that the mechanists were unable to acknowledge or to explain. But we no longer live in the age of mechanical determinism. Contemporary physics grew out of the breakdown of Newtonian mechanics at the beginning of the present century, both at the submolecular quantum domain and in the universe at large. We have as yet to work out the full implications of all this for biology. Some major thinkers early in the present century, such as the philosopher-mathematician, Alfred North Whitehead, already saw the need to explain physics in terms of a general theory of the organism[4], thus turning the usually accepted hierarchy of reductionistic explanation in science on its head. Whitehead's view is not accepted by everyone, but at least, it indicates that the traditional boundaries between the scientific disciplines can no longer be upheld, if one is to really understand nature. Today, physics has made further in-roads into the 'organic' domain, in its emphasis on nonlinear phenomena far from equilibrium, on coherence and cooperativity which are some of the hallmarks of living systems. The vitalist/mechanist opposition is of mere historical interest, for it is the very boundary between living and non-living that is the object of our enquiry, and so we can have no preconceived notion as to where it ought to be placed.

Similarly, to those of us who do not see our quest for knowledge as distinct from the rest of our life, there can be no permanent boundary between science and other ways of knowing. Knowledge is all of a piece. In particular, it is all of a piece with the knowing consciousness, so there can be no *a priori* dualism between consciousness and science. Far from implying that consciousness must be 'reduced' to physics and chemistry, I see physics and chemistry evolving more and more under the guidance of an active consciousness that *participates* in knowing[5]. Some of these issues will be dealt with in the last two chapters.

The Physicochemical Underpinnings of Life

Schrödinger's preliminary answer to the question of what is life is as follows:

"The obvious inability of present-day [1940's] physics and chemistry to account for such events [as take place within living organisms] is no reason at all for doubting that they can be accounted for by those sciences."[6]

He is saying that we simply do *not* know if events within living organisms could be accounted for by physics and chemistry because we have nothing like living systems that we could set up or test in the laboratory. There is a serious point here that impinges on the methods and technologies we use in science. Until quite recently, the typical way to study living organisms is to kill and fix them, or smash them up into pieces until nothing is left of the organization that we are supposed to be studying. This has merely reinforced the Newtonian mechanical view of organisms that has proved thoroughly inadequate to account for life. The situation is changing now with great advances in the development of non-invasive technologies within the past twenty years. We can 'listen in' to nature without violating her. I shall have more to say on that in later chapters.

Another reason for not doubting that physics and chemistry can one day account for living systems is surely that they are both evolving disciplines. Who knows what the subjects will look like in twenty year's time? Already, physics and chemistry today look quite different from the subjects half a century ago. The transistor radio, the computer and lasers have been invented since Schrödinger wrote his book. One major current obsession is with nano-technologies, or technologies which are precise to the molecular level because they make use of actual molecules. Whole new disciplines have been created: synergetics - the study of cooperative phenomena, nonequilibrium thermodynamics, quantum electrodynamics and quantum optics, to name but a few. In mathematics, non-linear dynamics and chaos theory took off in a big way during the 1960s and 70s. Perhaps partly on account of that, many nonlinear optical phenomena associated with quantum cavity electrodynamics and coherent light scattering in solid state systems are being actively investigated only within the past ten years. Meanwhile the race is on for the ultimate in high temperature superconducting material.

There have been several suggestions that these recent developments in physics and chemistry are particularly relevant for our understanding of biological phenomena. But there has not been a serious attempt to re-examine the issue in the way that Schrödinger has done in his time. I believe we have made substantial progress since, and it is the purpose of this book to substantiate that claim. By way of exploration, I shall range widely over equilibrium and nonequilibrium thermodynamics, aspects of quantum theory and solid state physics, as well as the relevant physiology, biochemistry and molecular biology of cells and organisms. I shall not be referring much to the details of molecular genetics and gene control mechanisms which already fill volumes of excellent texts, and which I simply take for granted. They are all part of the rich tapestry of life that will find their rightful place when our life-picture has been sufficiently roughed out. It is indeed the bare-bones sketch that I shall be concerned with here - the fundamental physical and chemical principles which make life possible.

I promise neither easy nor definitive answers. Our education already suffers from a surfeit of facile, simplistic answers which serve to explain away the phenomena, and hence to deaden the imagination and dull the intellect. An example is the claim that the natural selection of random mutations is necessary and sufficient to account for the evolution of life. As a result, whole generations of evolutionary biologists are lulled into thinking that any and every characteristic of organisms is to be 'explained' solely in terms of the 'selective advantage' it confers on the organism. There is no need to consider physiology or development, nor indeed the organism itself; much less the physical and chemical basis of living organization[7].

To me, science is a quest for the most intimate understanding of nature. It is not an industry set up for the purpose of validating existing theories and indoctrinating students in the correct ideologies. It is an adventure of the free, enquiring spirit which thrives not so much on answers as unanswered questions. It is the enigmas, the mysteries and paradoxes that take hold of the imagination, leading it on the most exquisite dance. I should be more than satisfied, if, at the end of this book, I have done no more than keep the big question alive.

What is life? Can life be defined? Each attempt at definition is bound to melt away, like the beautiful snowflake one tries to look at close-up. Indeed, there have been many attempts to define life, in order that the living may be neatly separated from the nonliving. But none has succeeded in capturing its essential nature. Out of the sheer necessity of communicating with my readers, I shall offer my own tentative definition, which to me, at least, seems closer to the mark: *life is a process of being an organizing whole*. It is important to emphasize that life is a *process* and not a thing, nor a property of a material thing or structure. As is well known, the material constituents of our body are continually being broken down and resynthesized at different rates, yet the whole remains recognizably the same individual being throughout. Life must therefore reside in the pattern of dynamic flow of matter and energy that somehow makes the organisms alive, enabling them to grow, develop and evolve. From this, one can see that the 'whole' does not refer to an isolated, monadic entity. On the contrary, it refers to a system open to the environment, that enstructures or organizes itself (and its environment) by simultaneously 'enfolding' the external environment and spontaneously 'unfolding' its potential into highly reproducible or dynamically stable forms[8]. To be alive and whole is a very special being. Let us dwell on that for a while.

On Being Alive

Biology textbooks often state that the most important characteristic of organisms is the ability to reproduce, and then proceed to give an account of

DNA replication and protein synthesis as though that were the solution to the fundamental problem of life. The ability to reproduce is only one of the properties of living organisms, and it could be argued, not even the most distinguishing one. For there are a number of other characteristics, scientifically speaking, which leave us in no doubt that they are alive: their extreme sensitivity to specific cues from the environment, their extraordinary efficiency and rapidity of energy transduction, their dynamic long range order and coordination, and ultimately, their wholeness and individuality[9].

For example, the eye is an exquisitely sensitive organ, which in some species, can detect a single quantum of light, or photon, falling on the retina. The photon is absorbed by a molecule of *rhodopsin*, the visual pigment situated in special membrane stacks in the outer segment of a rod-cell (Fig. 1.1). This results eventually in a nervous impulse coming out at the opposite end of the cell, the energy of which is at least a million times that contained in the original photon. The amplification of the incoming signal is in part well understood as a typical 'molecular cascade' of reactions: the specific receptor protein, *rhodopsin*, on absorbing a photon, activates many molecules of a second protein, *transducin*, each of which then activates a molecule of the enzyme *phosphodiesterase* to split many molecules of cyclic guanosine monophosphate or cGMP. The cGMP normally keeps sodium ion channels open in the otherwise impermeable cell membrane, whereas the split non-cyclic GMP cannot do so. The result is that the sodium channels close up, keeping sodium ions out of the cell and giving rise to an increased electrical polarization of the cell membrane - from about -40mV to -70mV, which is sufficient to initiate the nerve impulse[10].

Molecular cascades are common to all processes involved in signal transduction, and it is generally thought that one of their main functions is to amplify the signal. Let us examine the visual cascade reactions more closely. There are notable caveats in the account given in the last paragraph. For one thing, the component steps have time constants that are too large to account for the rapidity of visual perception in the central nervous system, which is of the order of 10^{-2}s. Thus, it takes 10^{-2}s just to activate *one* molecule of phosphodiesterase after photon absorption. Furthermore, much of the amplification is actually in the initial step, where the single photon-excited

Figure 1.1 Diagram of a light sensitive rod cell. The top part is the 'rod' containing membrane stacks in which the light sensitive pigments are situated. The resultant nervous impulse goes out at the bottom.

rhodopsin passes on the excitation to at least 500 molecules of transducin within one millisecond. How that is achieved is still a mystery, except that as rhodopsin and transducin molecules are bound to a common membrane, the membrane must play a crucial role in both the amplification and the long range transfer of excitation.

Another instructive example is muscle contraction[11]. Skeletal muscle consists of long, thin muscle fibres, several centimeters in length, each of which is a giant cell formed by the fusion of many separate cells. A single fibre, some 50μ (1μ = 10^{-6}m) in diameter, consists of a bundle of *myofibrils*, each 1 to 2 μ in diameter. A myofibril has regular, 2.5μ repeating units or *sarcomeres* along its length. Adjacent myofibrils are aligned so that their sarcomeres are in register, and electronmicrographs of fixed sections of muscle reveal extremely regular arrays of periodic structures (Fig 1. 2). (How these come about is just a small mystery in the whole process of development whereby a relatively featureless egg turns into a complicated, shapely and highly differentiated organism.) Each sarcomere consists of alternating thin and thick filaments made up respectively of polymers of *actin* complexed with other proteins, and *myosin*. The thin filaments are attached to an end-plate, the Z-disc (see Fig. 1. 2). Contraction occurs as the alternating myosin and actin fibres slide past each other by cyclical molecular treadmilling between myosin head groups and serial binding sites on the actin fibre. The sarcomere shortens proportionately as the muscle contracts.

Thus, when a myofibril containing a chain of 20,000 sarcomeres contracts from 5 to 4 cm, the length of each sarcomere decreases correspondingly from 2.5 to 2μ. The energy for contraction comes from the hydrolysis of the energy transacting intermediate, adenosine triphosphate, ATP, into adenosine diphosphate, ADP, and inorganic phosphate, Pi.

The sequence of events leading up to contraction begins with the firing of the nerve supplying the muscle, which triggers an *action potential* in the muscle-cell plasma membrane. An action potential is a quick electrical discharge followed by recovery of the pre-existing baseline electrical potential. The electrical excitation spreads rapidly into a series of membranous folds, the *transverse tubules*, that extend inwards from the plasma membrane to surround each myofibril at the region of the Z-disc. Here, the electrical signal is somehow transferred to a system of membrane-bound spaces, the *sarcoplasmic reticulum*, separate from the plasma membrane, and wrapped intimately around each myofibril. The sarcoplasmic reticulum then releases into the cell compartment containing the myofibril large amounts of Ca^{2+}. Careful measurements of the time course of Ca^{2+} release in muscle cells show that it begins almost immediately after electrical excitation, in other words, it takes hardly any time at all for the signal to traverse the cell[12]. The sudden rise in free Ca^{2+} initiates contraction simultaneously in the *entire cell*

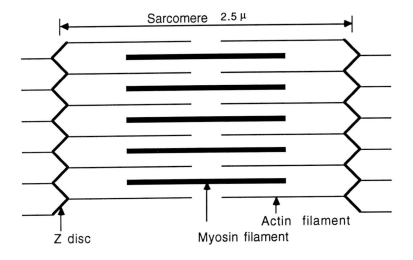

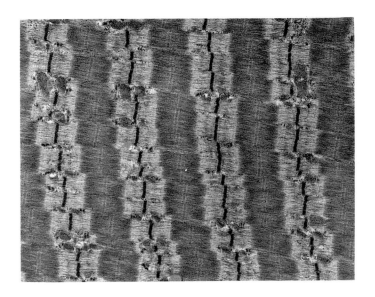

Figure 1.2 Ultrastructure of rabbit muscle. Top, diagram of a sarcomere; bottom, electronmicrograph.

within a millisecond. This involves numerous autonomously occurring cycles of attachment and detachment of all the individual myosin heads to and from the binding sites on the actin filaments at the rate of 50 cycles or more per second[13] - each of which molecular event requiring the transfer of energy contained in one molecule of ATP - precisely coordinated over the whole cell.

In a typical muscle contraction, all the cells (or fibres) in the muscle are executing the same molecular treadmilling, an astronomical number of them, in concert. This means that simply waving our arms about requires a series of actions coordinated instantaneously over a scale of distances spanning nine orders of magnitude from 10^{-9}m for intermolecular spacing to about 1m for the length of our arm; each action, furthermore, involving the coordinated splitting of 10^{20} individual molecules of ATP. Now, then, imagine what has to happen when a top athlete runs a mile in under four minutes! The same instantaneous coordination over macroscopic distances and astronomical number of molecules, only more so, and sustained for long periods without break.

It is truly remarkable how our energy should be available to us *at will*, whenever and wherever we want it, in the amount we need. Moreover, the energy is supplied at close to 100% efficiency. This is true for muscle contraction, in which the chemical energy stored in ATP is converted into mechanical energy[14], as well as for all the major energy transduction processes, as for example, in the synthesis of ATP itself in the mitochrondria[15] where carbon compounds are oxidized into carbon dioxide and water in the process of respiration. If that were not so, and energy transduction can only occur at the efficiency of a typical chemical reaction outside living organisms, which is 10 to 30% efficient at best, then we would literally burn out with all the heat generated.

Let us summarize what has been described. Being alive is to be extremely sensitive to specific cues in the environment, to transduce and amplify minute signals into definite actions. Being alive is to achieve the long range coordination of astronomical numbers of submicroscopic, molecular reactions over macroscopic distances; it is to be able to summon energy at will and to engage in extremely rapid and efficient energy transduction.

So, what constitutes this sensitive, vibrant whole that is the organism? An organism that, furthermore, develops from a relatively featureless fertilized egg or seed to a complicated shapely creature that is nonetheless the same essential whole?

We have certainly not exhausted the wonders of being alive, and shall continue our investigations from the standpoint of thermodynamics in the next four Chapters.

Notes

1. Schrödinger, (1944).

2. Schrödinger (1944). p. 3.

3. Needham (1935).

4. Whitehead, (1925).

5. I have written several papers recently on the need to recover a way of knowing in science that involves the complete participation of the knowing being: intellect and feeling, mind and body, spirit and intuition. Authentic science and art are both creative activities arising out of this total engagement of nature and reality. See Ho, (1989a; 1990a; 1993).

6. Schrödinger (1944), p. 4.

7. For alternative approaches to neo-Darwinism in the study of evolution, please see Ho and Saunders (1984); and Ho and Fox (1988).

8. See Ho (1988a).

9. See Ho (1989b).

10. See Stryer (1987).

11. See Alberts et al (1983).

12. Rios and Pizarro (1991).

13. See Pollard (1987).

14. Hibbard et al (1985).

15. Slater (1977).

CHAPTER TWO

DO ORGANISMS CONTRAVENE THE SECOND LAW?

Life and the Second Law

Many scientists have remarked that whereas the physical world runs down according to the second law of thermodynamics such that useful energy continually degrades into heat, or random molecular motion (expressed in the technical term, *entropy*), the biological world seems capable of doing just the opposite in increasing organization by a flow of energy and matter. Physicists and chemists have long felt that as all biological processes require either chemical energy or light energy and involve real chemical reactions, the second law, as much as the first law of thermodynamics (the conservation of energy) ought to apply to living systems. So, what is the secret of life? One explanation offered by Schrödinger[1] is that because living systems are open to the environment, they can create a local decrease in entropy at the expense of the surroundings, so that the entropy of living systems plus the surroundings always increases in all real processes, and there is no violation of the second law. But there are more fundamental problems, which Schrödinger was also aware of, as were the founding fathers of thermodynamics (see Chapters 5 and 11). We cannot appreciate those problems before we know what the laws of thermodynamics are, and to which systems they are supposed to apply.

What are the Laws of Thermodynamics?

Classical thermodynamics deals with the laws governing the conversion of heat into work or other forms of energy. It arose as a science giving an exact description of macroscopic systems of gases expanding against an external constraint such as a piston driving a steam engine, for example, for which the important parameters are pressure, volume and temperature. Its major foundations were laid long before detailed atomic theories became available. The possibility that the thermodynamic behaviour of matter could be derived from the mechanical properties (such as mass and velocity) of its constituent molecules forms the basis of *statistical mechanics*, which is supposed to give a rigorous deductive framework for thermodynamics[2].

There are two main laws of thermodynamics. The first is usually written as the equation,

$$\Delta U = Q + W \tag{1}$$

which says that the change in the total internal energy of a system, ΔU, is equal to the heat absorbed by the system from its surroundings, Q, plus the work done on the system by the surroundings, W. (The sign Δ - Greek letter *delta* - is shorthand for 'change in'.) It is based on the law of the conservation of energy, which states that energy is neither created nor destroyed in processes, but flows from one system to another, or is converted from one form to another, the amount of energy 'in the universe' remaining constant. The total internal energy, U, is a function of the particular state of the system, defined by temperature and pressure, and does not depend on how that state is arrived at. The heat absorbed, Q, and the work, W, are by contrast, *not* state functions and their magnitudes depend on the path taken. The first law tells us that energy is conserved but it does not tell us which processes can occur in reality. This is done by the second law.

The second law of thermodynamics tells us why processes in nature always go in one direction. Thus, heat flows from warm to cold bodies until the two are at the same temperature, but nobody has ever observed heat flowing spontaneously from cold to warm bodies. A drop of ink placed into a jar of water soon diffuses to fill the whole jar. And we would be surprised indeed, if, some time later, the original ink drop were to reconstitute itself. In the same way, machines run down unless constantly supplied with energy. Nobody has ever succeeded in building a perpetual motion machine which turns heat into mechanical work, then turns mechanical work back to its original equivalent in heat and so on again and again - which is not forbidden by the first law of thermodynamics.

Thus, spontaneous processes in nature define a time's arrow: they go in one direction only, never in reverse. Physical systems evolve from order to disorder, eventually running down until no more useful work can be extracted from them. To explain these phenomena, the second law defines a quantity, entropy, which increases in real processes and never decreases. Thus, at the end of a process, the change in entropy is always positive, or in the limiting case of a process which is reversible, the change in entropy is zero. This is expressed in the equation,

$$\Delta S_{sys} + \Delta S_{surr} \geq 0 \tag{2}$$

which says that sum of the changes in entropy in the system, ΔS_{sys}, and its surroundings, ΔS_{surr}, is greater than or equal to zero in all real processes. The change in entropy in each case is equal to the heat absorbed in a reversible process divided by the temperature at which the process occurs in degrees Kelvin, or K (which starts at -273.15°C, so that for example, 25°C would be 298.15K). For simplicity, we consider the process occurring in a system at constant temperature T,

$$\Delta S = Q_{rev}/T \tag{3}$$

A reversible process, in thermodynamic terms, is one that occurs very, very slowly, so that it is at equilibrium the whole way through. This means that the heat energy absorbed at every stage has time enough to spread or *equilibrate* throughout the system. In a reaction occurring reversibly, the net change of entropy in the system is exactly compensated by the change in entropy of the surrounding in the opposite direction, so that the net change in entropy is zero:

$$\Delta S_{sys} = -\Delta S_{surr}$$

$$\Delta S_{sys} + \Delta S_{surr} = 0$$

Another point about entropy is that it is a state function, and is therefore independent of the path taken to arrive at the particular state. If the process occurs irreversibly, the heat absorbed is less than that in the reversible process, but the entropy production does not change, so,

$$Q_{irrev}/T < \Delta S_{sys}$$

In that case, the required compensatory change in entropy in the environment cannot take place (on account of the reduction in the amount of heat lost to the system), and,

$$\Delta S_{sys} > -\Delta S_{surr}$$

$$\Delta S_{sys} + \Delta S_{surr} > 0$$

An example of an irreversible process is the expansion of a perfect gas against a vacuum. Here, no heat is absorbed from the surroundings and no work is done. The change in entropy of the system is the same as if the process occurred reversibly, but the change in entropy of the surroundings is zero, and so a net increase in entropy (so to speak, 'of the universe') results.

We can see that in the case of living systems, a *decrease* in entropy can indeed be achieved at the expense of a compensating increase in entropy in its surroundings, and there is no need to contravene the second law in that respect.

Another way to state the second law is that all isolated systems (those which exchange neither energy nor matter with their surroundings) run down so that useful energy is degraded into entropy. Entropy is thus made up of a kind of incoherent energy that is no longer available for work. (Entropy is, however, technically not the same as energy, as it has different dimensions, Joules per deg. K compared with Joules.) We can then define a number of functions for the free energy, or energy which is extractable for work: the Helmholtz free energy, A, and the Gibbs free energy, G,

$$A = U - TS$$

$$G = U + PV - TS$$

Thus, at constant temperature T, the change in free energies are,

$$\Delta A_{sys} = \Delta U_{sys} - T\Delta S_{sys} \qquad (4)$$

and,

$$\Delta G_{sys} = \Delta H_{sys} - T\Delta S_{sys} \qquad (5)$$

ΔH is the change in a quantity called *enthalpy*, or heat content, which is defined as,

$$H = U + PV$$

$$\Delta H = \Delta U + \Delta(PV) = \Delta U + P\Delta V + V\Delta P + \Delta P\Delta V$$

where U, P and V are the internal energy, pressure and volume of the system. At constant pressure, the last two terms disappear, and,

$$\Delta H = \Delta U + P\Delta V$$

In both expressions for free energy, we can see that as entropy increases, free energy decreases until at equilibrium, entropy reaches a maximum whereas free energy becomes a minimum.

The concept of thermodynamic equilibrium is central to the second law. It is a state of maximum entropy towards which all isolated systems evolve. A system at equilibrium is one in which no more changes can occur unless it is removed from isolation and placed in contact with another system.

So the laws of classical thermodynamics describe how one equilibrium replaces another. They do not say anything about the changes that happen in between when the system is not in equilibrium, and in that respect, are quite inadequate to deal with living systems where things are happening all the time (In Chapter 4, we shall see how thermodynamicists try to overcome that limitation with some success). A more serious problem is that the laws of thermodynamics, as usually formulated, apply to homogeneous bulk phase systems consisting of a large number of identical components[3].

Thus, entropy can be given an exact formulation in statistical mechanics by considering a large ensemble of identical systems at a given internal energy, volume and composition (a 'microcanonical ensemble' in the language of statistical mechanics), each of which can exist in a vast number of different microstates or more precisely, 'quantum states' (to be described in detail in Chapter 10). The entropy of the system is then given as,

$$S = k \, lnW \tag{6}$$

where the Boltzmann's constant, $k = 1.3805 \times 10^{-23}$ J K^{-1}, is a measure of the thermal energy in Joules associated with each molecule per degree K. W is the number of possible microstates that the system can exist in, and ln is the natural logarithm. (It should be apparent to you by now that physicists may use the same symbols for different entities: I have used W for work above for the first law of thermodynamics, whereas here it stands for microstates.) It can readily be appreciated that the greater the number of possible microstates, the greater the entropy, hence, of 'randomness'. The system consisting of the ink drop in a jar of water, for example, starts from a state of low entropy because the molecules are confined to a small region in the glass. Entropy increases as the molecules diffuse, so that finally, at equilibrium, a given molecule has the probability of being found in a small volume located anywhere in the glass. The equilibrium state is thus the one in which the entropy function given in eq. (6) is a maximum. Another way to say the same thing is that the equilibrium state is the most probable state.

Similarly, the total energy of a system corresponds to the sum of all the molecular energies of translation, rotation, vibration, plus electronic energy and nuclear energy (see Chapter 5); whereas temperature is proportional to the sum of the kinetic energies of all the molecules and, for a monatomic system, is defined by the formula:

$$\tfrac{3}{2}nkT = \sum_{i=1}^{n} \tfrac{1}{2}m_i C_i^2 \tag{7}$$

where n is the number of molecules, m_i is the mass of the ith molecule and C_i its velocity. The term kT is often referred to as the thermal energy of a molecule at temperature T.

The translation of macroscopic parameters into microscopic properties of molecules is by no means straightforward. For example, the statistical entropy function, elegant though it is, cannot easily be used for systems involving chemical reactions, which include all living systems as well as all chemical systems. It is only an analogue of the macroscopic entropy. As physical chemist, Kenneth Denbigh[4] reminds us, there is no *necessary* connection between entropy and either 'orderliness' or 'organization'. Along with concepts like energy, work, heat and temperature, entropy does not bear up to rigorous analysis[5]. We shall leave these difficulties aside for the moment (to be dealt with in Chapter 5), and concentrate on the limitations of the second law of thermodynamics which arise on account of its statistical foundations.

Is Maxwell's Demon in the Living System?

In his physical chemistry textbook, Glasstone[6] expresses a very commonly held view that the second law of thermodynamics is a statistical law which can only be applied to a system consisting of a large number of particles, and that, "If it were possible to work with systems of single, or a few, molecules, the law might well fail."

A major difficulty, already pointed out by Schrödinger, is that single molecules, or very few of them, are the active agents in living systems. Thus, each cell contains only one or two molecules of each sequence of DNA in the nucleus. Each bacterium of *E. coli* contains several molecules of the protein that enables it to respond to the presence of lactose in its environment, resulting in the induction of several enzymes involved in metabolizing the sugar and enabling it to grow and to multiply. This is typical of whole classes of metabolic regulators. Similarly, it takes no more than several molecules of a given hormone to bind to specific receptors in the cell membrane in order to initiate a cascade of biochemical reactions that alter the characteristics of the whole cell. Does this mean the second law cannot be applied to living systems?

Actually, this difficulty is not restricted to biology, but occurs in physical systems as well. The most colourful statement of the problem is in the form of Maxwell's demon[7] - an hypothetical intelligent being who can operate a microscopic trapdoor between two compartments of a container of gas at equilibrium so as to let fast molecules through in one direction, and the slow ones in the other. Soon, a temperature difference would be created by the accumulation of fast, energetic molecules on one side and slow ones on the other, and work can then be extracted from the system. Maxwell invented this demon in 1867 to illustrate his belief that the second law is statistical, and had no intention of questioning the second law itself. The trapdoors, after all, would also be subject to the same statistical fluctuations as the molecules, and would open and close indiscriminately so that the separation of fast from slow molecules could never be achieved, unless we had the magical demon - small and clever enough to observe the fluctuations.

Thirty-eight years later, however, Einstein showed that the fluctuations *can* be observed, and in fact, were first observed in 1827 by Robert Brown - as Brownian motion. This is the random movements of microscopic particles (observable under the microscope) as they are jostled about by the fluctuations of the water molecules. It also became evident in the 1950s that something like a Maxwell's demon could be achieved with little more than a trapdoor that opens in one direction only and requires a threshold amount of energy (activation energy) to open it. This is realizable in solid-state devices such as diodes and transistors that act as rectifiers[7]. Rectifiers let current pass in one direction only but not in reverse, thereby converting alternating currents to direct currents. This means that they can convert randomly fluctuating currents, in principle, into an electrical potential difference between the two sides from which work can then be extracted.

Similar situations are commonly found in biological membranes which have electrical potential gradients of some $10^7 V/m$ across them and have bound enzymes involved in the vectorial transport of ions and metabolites from one side to the other, as for example, the transport of Na^+ out of, and

K^+ into the cell by the Na^+/K^+ ATPase. It has recently been demonstrated that weak alternating electric fields can drive unidirectional active transport by this enzyme without ATP being broken down. In other words, the energy from the electric field is directly transduced into transport work by means of the membrane-bound enzyme. Moreover, randomly fluctuating electric fields are also effective, precisely as if Maxwell's demon were involved in making good use of the fluctuations[8]!

The problem of Maxwell's demon is generally considered as having been 'solved' by Szilard, and later, Brillouin, who showed that the demon would require information about the molecules, in which case, the energy involved in obtaining information would be greater than that gained and so the second law remains inviolate. What they have failed to take account of is that the so-called information is already supplied by the special structure or organization of the system (see Chapter 11). In the next chapter, I will concentrate on how the problem of Maxwell's demon might be solved.

Notes

1. Schrödinger (1944).

2. See Penrose (1970).

3. Statistical mechanics can be formulated to deal with a mixture of chemical species, and to a limited extent, with space-time structure, but not yet of the kind that exists in living systems. I am very grateful to Geoffrey Sewell for pointing this out to me.

4. See Denbigh (1989), pp. 323-32.

5. See Bridgman, (1941). Geoffrey Sewell disagrees with my statement, pointing out that a lot of effort has since been devoted to translating/relating macroscopic entropy to degree of molecular disorder either in a system (von Neumann entropy) or in a process (Kolmogorov entorpy). However, I believe it is still true to say that the relationship between the macroscopic and microscopic entities remain conjectural. There is no *necessary* logical connection between the two (see Chapter 5).

6. Glasstone (1955).

7. Ehrenberg (1967).

8. See Astumian *et al* (1989). The authors are at pains to point out that noise internal to the system, however, cannot be used in the same way to generate work, because electrical correlations would be induced to oppose it, and hence the second law is by no means violated.

CHAPTER THREE

CAN THE SECOND LAW COPE WITH ORGANIZED COMPLEXITY?

The Space-time Structure of Living Processes

One cannot fully appreciate the problem of Maxwell's demon in the context of living organisms without taking full account of the complexity of the organism's space-time structure, which demonstrates clearly the limitations of thermodynamics applied to living systems.

To begin, we can take it as obvious and given that the organism is an open system, which moreover, qualifies as a *dissipative structure*[1] in the sense that its organization is maintained in some kind of 'steady state' by a flow of energy and chemicals. (We shall say more about energy flow in the next chapter.) As soon as that flow is interrupted, disintegration sets in and death begins. That steady state, however, is not a static bulk phase in a rigid container such as the physical chemists's continuously stirred tank reactor (CSTR) or, as microbiologists prefer to call it, the *chemostat*. The CSTR has long served as the model for the steady state, and a lot of useful analyses have been achieved. But as far as a representation of living organization is concerned it introduces some quite misleading features. Indeed, in typical models of dissipative structures such as the Bénard convection cells (see next chapter) which develop in a shallow pan of water heated from below, or the famous Beloussov-Zhabotinsky oxidation-reduction reaction, which gives oscillating, concentric red and blue rings and various spiralling patterns in a petri dish (see Fig. 3.1), the dynamical structures - the objects of interest - are obtained precisely because the system is not stirred. Stirring would obliterate those structures and featureless homogeneity would result[2].

What do we find in the organism? Organized heterogeneities, or dynamic structures on all scales. There is no homogeneity, no static phase held at any

22

level. Even a single cell has its characteristic shape and anatomy, all parts of which are in constant motion; its electrical potentials and mechanical properties similarly, are subject to cyclic and non-cyclic changes as it responds to and counteracts environmental fluctuations. Spatially, the cell is

Figure 3.1 The Beloussov-Zhabotinsky reaction[3].

partitioned into numerous compartments by cellular membrane stacks and organelles, each with its own 'steady states' of processes that can respond directly to external stimuli and relay signals to other compartments of the cell. Within each compartment, microdomains[4] can be separately energized to give local circuits, and complexes of two or more molecules can function as 'molecular machines' which can cycle autonomously without immediate reference to its surroundings. In other words, the steady 'state' is not a state at all but a conglomeration of processes which are spatiotemporally organized, i.e., it has a deep space-time structure, and cannot be represented as an instantaneous state or even a configuration of states[5]. Characteristic times of processes range from $<10^{-14}$ s for resonant energy transfer between molecules to 10^7 s for circannual rhythms. The spatial extent of processes, similarly, spans at least ten orders of magnitude from 10^{-10} m for intramolecular interactions to metres for nerve conduction and the general coordination of movements in larger animals.

The processes are catenated in both time and space: the extremely rapid transient flows (very short-lived pulses of chemicals or of energy) which take place on receiving specific signals, are propagated to longer and longer time domains of minutes, hours, days, and so on via interlocking processes which ultimately straddle generations. These processes include the by now familiar enzyme activation cascades (see Chapter 1) which occur in response to specific stimuli, and often end in the expression of different genes and in morphological differentiation.

For example, repeated stimulation of the same muscles encourages the growth of those muscles and make them function more efficiently, as body builders are well aware! The intermediate events include changes in innervation, and the expression of genes which code for different sets of muscle proteins[6]. The graph in Figure 3. 2 depicts the catenated sequence of events and their approximate timescales. A transient pulse of a chemical signal, *acetylcholine*, sent out by the nerve cell at its junction with the muscle, opens the sodium ion channels in the membrane of the muscle cell and depolarizes the cell membrane within 10^{-3}s (peak 1), triggering an influx of Ca^{2+} from the sarcoplasmic reticulum which lasts for 10^{-2}s (peak 2). This in turn sets off the reactions between actin and myosin in a contraction, which involves many cycles of molecular treadmilling where ATP is split into ADP and Pi. Each individual cycle is some 10^{-2} to 10^{-1}s long (peak 3), whereas a contraction may last 1 to 10s (peak 4). Sustained muscular activities, consisting of numerous cycles of contraction and relaxation of entire muscles, go on typically for 10^2 - 10^3s (peak 5). This stimulates the transcription of specific genes to increase the synthesis of special muscle proteins in 10^3s or longer (peak 6). Over a period of days, or months, repetition of the same sequence of activities in regular exercises gives rise to the desired changes to the muscles involved (peak 7): enhancement in anatomy and improvement in performance.

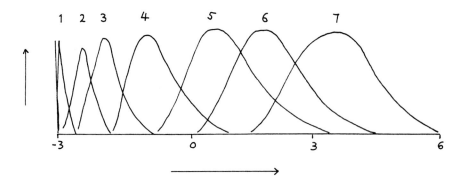

Figure 3.2 The catenation of living processes in time. The vertical axis represents the magnitude, or amplitude, of change; the horizontal axis is time in seconds in logarithm to the base ten. The amplitudes are represented as equal for all the processes, but in reality, they may be progressively amplified or diminished at longer and longer time scales see text.).

These catenated processes are responsible for the phenomenon of 'memory' so characteristic of living systems. In reality, the processes are projections or propagations into the *future* at every stage. They set up influences to determine how the system develops and responds in times to come. Typically, multiple series of activities emanate from the focus of excitation. The mere *anticipation* of muscular activity is accompanied by the secretion of adrenaline, which in turn causes the blood vessels to dilate, increasing the heart rate, and thus enhancing the aeration of muscles and the synthesis of more ATP to supply the sustained activity of the muscle. While the array of changes in the positive direction is propagating, a series of negative feedback processes is also spreading, which has the effect of dampening the change. It is necessary to think of all these processes propagating and catenating in parallel in many dimensions of space and time (see Fig. 3.3). In case of disturbances which have no special significance for the body, homeostasis is restored sooner or later as the disturbance passes. On the other hand, if the disturbance or signal is significant enough, a series of irreversible events brings the organism to a new 'steady state' by developing or differentiating new tissues. The organism may even act to alter its environment appropriately.

The dynamism of the living system is such that each single cell in the body is simultaneously criss-crossed by many circuits of flow, each circuit with its own time domain and direction, specified by local pumping, gating and chemical transformation. Thus, classical equilibrium constants are quite irrelevant[4] for each 'constant' is in reality a continuous function of variables including the flow rates, the electrical and mechanical field strengths and so on. Furthermore, as the reaction products change the chemical potentials of all the components by altering the variables, the equilibrium 'constants' will also be functions of time. How can we describe such a space-time structure? The inorganic chemist, R.J.P. Williams[4], who has considered this problem with the refreshing eye of 'first looking into nature's chemistry', advocates a shift from a conventional thermodynamic approach to a dynamic approach, which would involve a description of the living system in terms of forces and flows rather than a succession of equilibrium states. This has already begun, to some extent, with nonequilibrium thermodynamics (see next chapter). But the fundamental difficulty of the statistical nature of a thermodynamic description remains practically untouched. There is as yet no science of organized heterogeneity or complexity such as would apply to living systems.

As is clear from the description above, living systems consist of compartments of various sizes down to microdomains each with its own steady state ; and complexes of a few molecules can act autonomously as

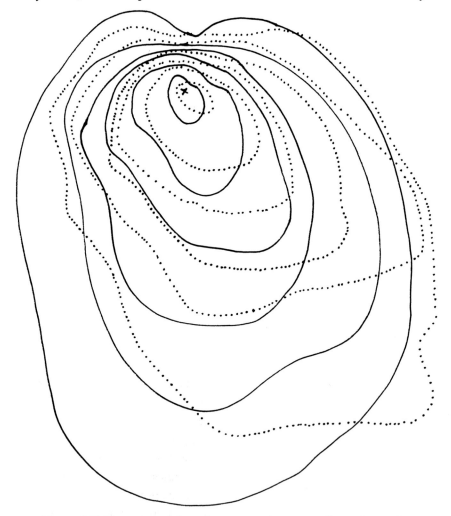

Figure 3.3 The propagating of processes in many dimensions of space and time. The spot marked x is the initial signalling process. The dotted lines and solid lines represent two series of catenated processes spreading out from the initial event.

efficient cyclic molecular machines. At the very least, this implies that if thermodynamics were to apply to living systems, it must apply to individual molecules. Such is the physiologist Colin McClare's[7] contention.

The Second Law Restated

In order to formulate the second law of thermodynamics so that it applies to single molecules, McClare introduces the important notion of a characteristic time interval, τ, within which a system reaches equilibrium at temperature θ. The energies contained in the sytem can be partitioned into *stored* energies versus *thermal* energies. Thermal energies are those that exchange with each other and reach equilibrium in a time less than τ (so technically they give the so-called Boltzmann distribution characterized by the temperature θ). Stored energies are those that remain in a non-equilibrium distribution for a time greater than τ, either as characterized by a higher temperature, or such that states of higher energy are more populated than states of lower energy. So, stored energy is any form which does not thermalize, or degrade into heat in the interval τ. Stored energy is *not* the same as free energy, as the latter concept does not involve any notion of time. Stored energy is hence a more precise concept.

McClare goes on to restate the second law as follows: useful work is only done by a molecular system when one form of stored energy is converted into another. In other words, thermalized energy is unavailable for work and it is impossible to convert thermalized energy into stored energy.

McClare is right in identifying the problem of Maxwell's demon in relation to the living system, and in stressing that useful work can be done by a molecular system via a direct transfer of stored energy *without thermalization*. The significance of this alone requires much more thought, as photosynthesis involves the direct, nonthermal absorption of the energy of photons, and non-thermal energy transfer may play a much larger role in living processes than hitherto recognized. (More on this is deferred until Chapter 7.) However, his restatement of the second law is unnecessarily restrictive, and possibly untrue, for thermal energy *can* be directed or channelled to do useful work in a cooperative system, as in the case of enzymes embedded in a membrane described at the end of the previous chapter. Thermalized energy from the burning of coal or petrol is routinely used to run machines such as generators and motor cars (which is why they are so inefficient and polluting). In the case of the motor car, the hot gases expand against the external constraint of the piston, which converts thermalized energy into mechanical work. McClare's difficulty in envisaging how thermalized energy could be used in the living system is at least in part due to the lack of such a constraint against which thermalized energy could be converted into useful work[8].

A more adequate restatement of the second law, I suggest, might be as follows:

Useful work can be done by molecules by a direct transfer of stored energy, and thermalized energy cannot be converted into stored energy.

The second half of the statement probably accounts for entropic decay as is usual in real processes both inside and outside the living system. The first half, however, is new and significant for biology, as it may be for the non-equilibrium phase transitions associated with laser action, for example (see Chapter 7).

The major consequence of McClare's ideas arises from the explicit introduction of time, and hence time-structure. For there are now two quite distinct ways of doing useful work, not only slowly according to conventional thermodynamic theory, but also quickly - both of which are reversible and at maximum efficiency as no entropy is generated. This is implicit in the classical formulation, $dS \geq 0$, for which the limiting case is $dS=0$, as explained in the previous chapter. But the attention to time-structure makes much more precise what the limiting conditions are. Let us take the slow process first. A slow process is one that occurs at or near equilibrium. According to classical thermodynamics, a process occurring at or near equilibrium is reversible, and is the most efficient in terms of generating the maximum amount of work and the least amount of entropy (see p.14). By taking explicit account of characteristic time, a reversible thermodynamic process merely needs to be slow enough for all thermally-exchanging energies to equilibrate, ie, slower than τ, which can in reality be a very short period of time, for processes that have a short time constant. Thus, for a process that takes place in 10^{-12}s, a millisecond (10^{-3}s) is an eternity! Yet for the cell in which it occurs, it is about the 'normal' timescale, and for us, a millisecond is an order of magnitude below the level of our time awareness. So high efficiencies of energy conversion can still be attained in thermodynamic processes which occur quite rapidly, provided that equilibration is fast enough. This may be where spatial partitioning and the establishment of microdomains are crucial for restricting the volume within which equilibration occurs, thus reducing the equilibration time. This means that *local equilibrium may be achieved at least for some biochemical reactions in the living system.* We begin to see that thermodynamic equilibrium itself is a subtle concept, depending on the level of resolution of time and space. I shall have more to say on that in the next chapter.

At the other extreme, there can also be a process occurring so quickly that it, too, is reversible. In other words, provided the exchanging energies are not thermal energies in the first place, but remain stored, then the process is limited only by the speed of light. Resonant energy transfer between molecules is an example of a fast process. As is well known, chemical bonds when excited, will vibrate at characteristic frequencies, and any two or more bonds which have the same intrinsic frequency of vibration will resonate with one another. (This happens also in macroscopic systems, as when a tuning fork is struck near a piano, the appropriate string will begin to vibrate when it is in tune.) More importantly, the energy of vibration can be transferred through large distances, theoretically infinite, if the energy is radiated, as electromagnetic radiations travel through space at the speed of light, though in practice, it may be limited by nonspecific absorption in the intervening medium. Resonant energy transfer occurs typically in 10^{-14}s, whereas the vibrations themselves die down, or thermalize, in 10^{-9}s to 10^{1}s. (On our characteristic time-scale - roughly 10^{-2}s - the vibrations would persist for as long as one year to one thousand years!) It is 100% efficient and highly specific, being determined by the frequency of the vibration itself; and resonating molecules (like people) can attract one another. By contrast, conventional chemical reactions depend on energy transfer that occurs only at collision, it is inefficient because a lot of the energy is dissipated as heat, and specificity is low, for non-reactive species could collide with each other as often as reactive species.

Does resonant energy transfer occur in the living system? McClare[7] suggests it occurs in muscle contraction, where it has already been shown that the energy released in the hydrolysis of ATP is almost completely converted into mechanical energy in a molecular machine which can cycle autonomously without equilibration with its environment (see previous Sec.). Similar cyclic molecular machines are involved in other major energy transduction processes: in the coupled electron transport and ATP synthesis in oxidative phosphorylation and photophosphorylation, as well as in the Na^+/K^+ ATPase. Ultrafast, possibly resonant energy transfer processes are also operating in photosynthesis[9]. There, the first step is the separation of positive and negative charges in the chlorophyll molecules of the reaction centre, which has been identified to be a readily reversible reaction that takes place in less than 10^{-13}s.

Thus, the living system may use both means of efficient energy transfer: slow and quick reactions, always with respect to the relaxation time, which is itself a variable according to the processes and the spatial extents involved. In other words, it satisfies both quasi-equilibrium and far from equilibrium conditions where entropy production is minimum[9]. This insight is offered by taking into account the space-time structure of living systems explicitly. We shall have the occasion to return again and again to entropy production and space-time structure in the living system in later chapters.

Quantum Molecular Machines in Living Systems

Another important insight is the fundamental *quantum* nature of biological processes. McClare[7] defines a molecular energy machine as one in which the energy stored in single molecules is released in a specific molecular form and then converted into another specific form so quickly that it never has time to become heat. It is also a quantum machine because it sums the effects produced by single molecules. Muscle contraction is the most obvious example, as described in Chapter 1.

Even in conventional enzyme kinetics, more and more quantum mechanical effects are recognized. Electron tunnelling (going 'under' an energy barrier by virtue of an overlap of the quantum mechanical wave functions of the electron in different neighbouring states) is already well known to be involved in the separation of charges and electron transport across the biological membranes of the chloroplasts of green plants as well as across electron transport proteins such as *cytochrome c.* Hydrogen transfer reactions may also involve tunnelling across energy barriers via an overlap of quantum mechanical wave functions between enzyme complexes of substrates and products. It may be that very few reactions occurring in organisms involve thermalization of stored molecular energy[8]. This does not preclude thermal excitation where the activation energy barrier is sufficiently low, as for example, in the making and breaking of the hydrogen bonds involved in maintaining the three-dimensional shapes or *conformations* of protein molecules (see Chapter 6). Indeed, such fluctuational changes in conformation occur within nanoseconds (10^{-9}s), and they have been observed in a large number of proteins. But, in order to do useful work, the fluctuations have to be coordinated. Otherwise there will be equal probability for the reaction to go forwards as backwards - precisely as predicted in statistical mechanics.

There seems to be no escape from the fundamental problem of biological organization: *how can individual quantum molecular machines function in collective modes extending over macroscopic distances?* Just as bulk phase thermodynamics is inapplicable to the living system, so perforce, some new principle is required for the coordination of quantum molecular machines. This principle is *coherence*, perhaps even *quantum coherence*; but we shall leave that for later. In the next chapter, we consider the consequences of energy flow in biology in greater detail.

Notes

1. This term has been invented by Nobel laureate physical chemist Ilya Prigogine in order to capture what he believes to be the essential nature of the living system that is shared with certain non-equilibrium physical and chemical systems. See Prigogine (1967).

2. Physical chemist, Michael Menzinger and his co-workers, have recently shown that even stirred reactors can maintain a variety of dynamic oscillatory structures. It seems clear that coupling between flow and chemical reaction plays a crucial role in generating structural complexities. See Menzinger and Dutt, (1990).

3. From Winfree and Strogatz (1983) p.37.

4. This is particularly emphasized by inorganic chemist, R.J.P Williams, who has taken a particularly fresh and inspiring look at living organization for us. See Williams (1980).

5. See Ho (1993).

6. See Lomo *et al* (1985); Laing *et al* (1985).

7. McClare (1971).

8. Denbigh, K. (personal communication) comments as follows: "..the applicability, or the non-applicability of McClare's theory can best be grasped by using $S = k \ln W$. In any proposed application, one asks: will the number of quantum states, W, increase or not? If the answer appears to be no, then reversibility can be assumed, and entropy remains constant. Otherwise, entropy increases."

9. Nonthermal energy transduction in living organisms has recently been reviewed by Ho and Popp (1993).

CHAPTER FOUR

ENERGY FLOW AND LIVING CYCLES

The Probability of Life

The chemical constituents of a typical living cell, say, an *E. coli* bacterium, are known in considerable detail. The major elements are carbon, hydrogen, nitrogen, oxygen, phosphorous and sulphur, in that order, or CHNOPS for short, making a sound like a cow munching on fresh, green grass. These go to make up the organic polymers such as proteins, nucleic acids, membrane lipids, carbohydrates, and small molecular weight cofactors and intermediates. A question of great interest to scientists is: how probable is life in its simplest form of the cell? The problem can be considered as follows. Suppose one were to mix together the chemical elements in their appropriate amounts and proportions in each of an infinite array of sealed containers kept indefinitely in a very big water bath (or reservoir) at 300°C, what fraction of them would eventually develop into living cells? The answer is $10^{-10^{11}}$- an infinitesimally small number, so small that there has not been time, nor matter enough in the whole of the universe for it to happen even once. How is such a number arrived at? Is there any basis to this kind of estimate?

The (hypothetically) infinite array of systems with the same chemical composition, volume and temperature, is a 'canonical ensemble', in the language of statistical mechanics, which is allowed to go to equilibrium. The theory is that eventually, *every* possible configuration of the atoms or microstates, i, with corresponding energy level, e_i, will be explored, some of which would correspond to those in the living system. But because the energy level of the living state is so much higher than the average in the equilibrium ensemble, the probability of those states occurring is vanishingly small.

The living cell has a very large amount of energy stored in covalent bonds as electronic bond energies - considerably more than the thermal energies which exist in the equilibrium state. So large, in fact, that the probability of getting there by chance fluctuation around the equilibrium state is essentially nil. So how come there are living organisms at all? The answer is energy flow. Energy flow greatly increases the probability for life, and is absolutely essential for its maintenance and organization. These are some of the insights offered by physical chemist, Harold Morowitz, in his wonderful book, *Energy Flow in Biology*[1]. We shall explore some of his ideas in the pages following.

Energy Flow and Material Cycles

The concept of energy flow in biology is familiar to every biochemist and ecologist. It is the energy from sunlight, trapped by green plants, which flows through the whole biosphere, beginning with herbivores and insects which feed directly on green plants, through to other 'trophic layers' of animals that in turn feed upon them. At every stage, the energy is further transformed, rearranged and degraded, and a fraction lost as heat. Accompanying the energy flow is a *cycling* of materials through the biosphere: carbon dioxide, water, nitrogen, sulphates and phosphates are 'fixed' by green plants and bacteria into high electronic energy compounds which cycle through the bodies of animals in the biosphere before they are broken down and returned to the soil and the atmosphere for general recycling. (Cycling and re-cycling are indeed nature's way.) The relationship between energy flow and the cycling of the elements in the biosphere is represented very schematically in Figure 4.1. The 'cycle' is actually a lot more complicated. It is a composite of many parallel cycles and epicycles variously connected, and branching at different points.

This intimate relationship between energy flow and the cycling of the elements turns out not to be at all fortuitous. For it is the energy flow which organizes the material system, taking it far away from thermodynamic equilibrium by raising its energy level, producing gradients, and cyclic flow patterns of materials (and energy), in short, generating a hierarchy of space-time structures which in turn organizes the energy flow. The key to life is in this mutuality of spontaneous relationship between the system and the environment, each in turn the organizer and the organized (Fig. 4.2).

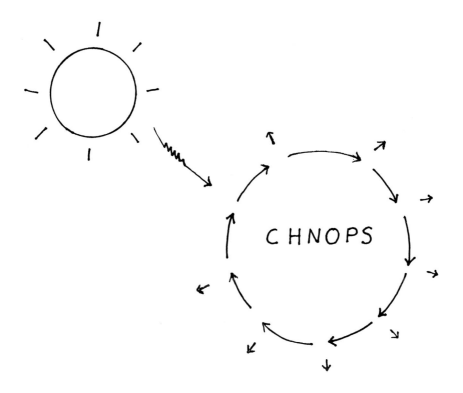

Figure 4.1 The CHNOPS cycle and energy flow.

The same space-time catenation of processes occurs on the ecological level as in individual organisms, involving larger dimensions and longer durations, of course. Ecological processes are continuous with, and impinge upon individual organismic processes, and are subject to similar cybernetic principles of regulation; so much so that 'geophysiologist', James Lovelock, has proposed that the earth is like a superorganism[2]. And hence, a lot could be learned by concentrating on the global homeostatic, feedback mechanisms that account for the stability of the earth's ecological system as a whole. Many other scientists, including Sidney Fox[3] who works on the origin of life, also believe that biological evolution is continuous with the evolution of the solar system and chemical evolution, and is by no means the result of a series of lucky 'frozen accidents'.

Energy Flow

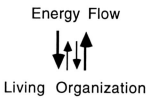

Living Organization

Figure 4.2 Energy flow and living organization.

These fascinating aspects are beyond the scope of the present book, but they do emphasize the continuity between the living and non-living, which *is* my thesis. Without further ado, let us go on to consider Morowitz's idea that energy flow leads to space-time structures, in particular, cycles.

Dynamic Molecular Order from Energy Flow

Some simple examples will illustrate how molecular order can arise from energy flow. Figure 4.3 is a model system consisting of a chamber containing an ideal gas (ie, at low enough concentrations such that the volume of individual molecules do not matter, and the individual molecules do not interfere much with one another). The left side is kept indefinitely at temperature T_1 by contact with a large reservoir at the same temperature, and the right side kept by contact with an equally large reservoir, at temperature

T_2; with $T_1 > T_2$. A steady flow of heat is therefore maintained across the chamber from left to right, or from 'source' to 'sink'.

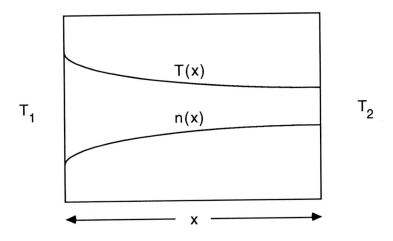

Figure 4.3 Heat flow and the formation of a concentration gradient in a volume of ideal gas.

The result is a temperature gradient from left to right, and at the same time, a concentration gradient of the molecules increasing from the 'hot' end to the 'cold' end. This has been theoretically predicted and experimentally observed. The calculations are quite complicated, but it is intuitively easy to see how the concentration gradient arises. The hotter molecules at the left have more kinetic energy (which is proportional to kT, see p. 18); therefore, they tend to have higher velocities than those on the right, and there is a net (convectional) movement of molecules towards the right where they cool down and lose kinetic energy. At steady state, this is balanced by diffusional movement from the right back to the left[4]. This would go on indefinitely as long as the temperature difference is maintained. If now the chamber is 'adiabatically' isolated from the reservoirs, ie, isolated so that there is no heat exchange with the reservoirs, it would soon decay back to uniform equilibrium in which the temperature and concentration everywhere will be the same. This is the simplest version of a 'dissipative structure'[5] which must be maintained by a continuous flow, or dissipation, of energy. The next example involves the formation of more complicated structures.

A shallow pan of water is heated uniformly from below (Fig. 4.4). For a small temperature difference, the situation is similar to the previous example, a temperature gradient is maintained across the depth of the water from the bottom to the top, and a density gradient in the opposite direction. The one difference here is that the heat transfer will occur by conduction rather than convection, for the water molecules themselves do not alter their mean position in the process.

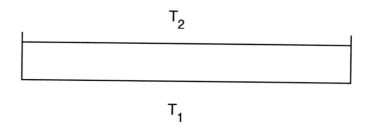

T_2

T_1

Figure 4.4 Heat flow in a shallow pan of water.

If we now increase the temperature difference, a critical value will be reached at which bulk movements will occur in the liquid. A series of regular, convection cells are formed, giving a honey-comb appearance when viewed from above (Fig. 4.5). These so-called Bénard convection cells arise as the low density, lighter water at the bottom of the pan repeatedly rises to the top while the denser water at the top sinks to the bottom, and so on in a cyclic manner. The detailed mathematical analysis is beyond the scope of this book, but it is possible to imagine that as convectional movements begin in different parts of the pan and increase in magnitude, the system would soon end up in a situation where all the convection cells are of the same size and are cycling together, as that is the most stable dynamical state. Let us dwell on this remarkable state in terms of what is happening at the molecular level. Each cell involves the concerted cyclical movement of some 10^{23} individual water molecules (as schematically represented in Fig. 4.6)[6]. Furthermore, all the cells in the pan are synchronized with respect to one another. In technical language, *a non-equilibrium phase transition to macroscopic order has taken place.* Before phase transition, the individual water molecules

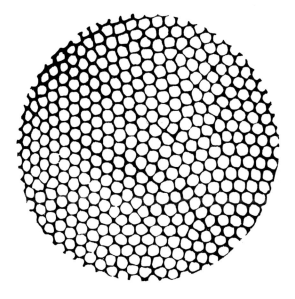

Figure 4.5 The Bénard convection cells[6].

move around randomly and aimlessly without any order or pattern. At phase transition, however, they begin to move cooperatively until all the molecules are dancing together in cellular formations as though choreographed to do so. Like the previous example, Bénard convection cells are dissipative structures dependent on the flow of energy for their continued existence. It is appropriate also to refer to them as 'coherent structures' which are dynamically maintained. Do keep in mind this wonderfully evocative picture of global order constituted of microscopically coherent motions when we try to imagine what happens in a living system later on.

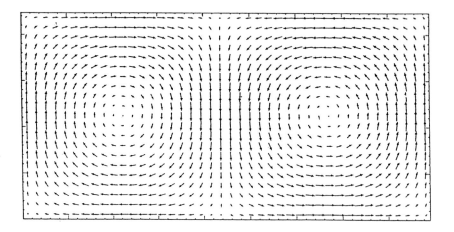

Figure 4.6 Coherent molecular motions in the convection cells[6].

In both examples of nonequilibrium structures just described, there is a net cycling of molecules around the system, from the hot end to the cold end and back again, the situation being more dramatic in the second case. What happens when chemical reactions are involved? Does chemical cycling also take place as the result of energy flow?

A Theorem of Chemical Cycles

Consider a reactor maintained by contact with a large reservoir at temperature T, in which are three chemical species, A, B, and C, reacting according to the following scheme:[7]

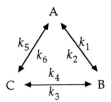

where the k's are the rate constants: the odd numbered ones refer to rates in the forward direction - A to B to C, and the even numbered ones in the reverse direction. At equilibrium, ie, without energy flow, forward and backward reactions are equal. This is *the principle of microscopic reversibility* or *detailed balance*, which is of great importance in chemistry. Hence,

$$
\begin{aligned}
k_1[A] &= k_2[B] \\
k_3[B] &= k_4[C] \\
k_5[C] &= k_6[A]
\end{aligned}
\tag{1}
$$

where the bracketed letters represent concentrations of the respective chemical species. Thus, at equilibrium, there is no net flow around the system, as the forward and backward reactions are occurring at the same rate and are therefore in detailed balance at every step. Under the additional constraint that the total amount of material in the reactor is constant, ie,

$$
[A] + [B] + [C] = M
\tag{2}
$$

we can solve for the equilibrium concentrations of each species. These are,

$$
\begin{aligned}
[A]_{eq} &= k_2 k_4 M / k' \\
[B]_{eq} &= k_1 k_4 M / k' \\
[C]_{eq} &= k_1 k_3 M / k'
\end{aligned}
\tag{3}
$$

where $k' = k_2 k_4 + k_1 k_4 + k_1 k_3$.

If energy now flows into the system so that a steady state is maintained with a net flow of material around the system from A to B to C, the principle of microscopic reversibility will no longer hold: there will not be a detailed balance in every step. In the three species system,

$$k_1[A] > k_2[B]$$
$$k_3[B] > k_4[C] \tag{4}$$
$$k_5[C] > k_6[A]$$

In other words, the material must cycle around the system. The flow, or rate at which the material is cycling around the system is given by,

$$F = k_1[[A] - k_2[B] = k_3[B] - k_4[C] = k_5[C] - k_6[A] \tag{5}$$

This result can be restated in a formal, general way. A canonical ensemble of systems is at equilibrium with i possible states, where f_i is the fraction of systems in state i (also referred to as occupation numbers of the state i), and t_{ij} is the transition probability that a system in state i will change to state j in unit time. The principle of microscopic reversibility requires that every forward transition is balanced in detail by its reverse transition, ie,

$$f_i t_{ij} = f_j t_{ji} \tag{6}$$

If the equilibrium system is now irradiated by a constant flux of electromagnetic radiation such that there is net absorption of photons by the system, a steady state will be reached at which there is a flow of heat out into the reservoir (sink) equal to the flux of electromagnetic energy into the system. At this point, there will be a different set of occupation numbers and transition probabilities, f_i' and t_{ij}'; for there are now both radiation induced transitions as well as the random thermally induced transitions characteristic of the previous equilibrium state. This means that for some pairs of states i and j,

$$f_i' t_{ij}' \neq f_j' t_{ji}' \tag{7}$$

For, if the equality holds in all pairs of states, it must imply that for every transition involving the absorption of photons, a reverse transition will take place involving the radiation of the photon such that there is no net absorption of electromagnetic radiation by the system. This contradicts our original assumption that there is absorption of radiant energy (see previous paragraph), so we must conclude that the equality of forward and reverse transitions does not hold for some pairs of states. However, at steady state, the occupation numbers (or the concentrations of chemical species) are time independent (ie, they remain constant), which means that the *sum* of all forward transitions equals to the *sum* of all backward transitions, ie,

$$df_i'/dt = 0 = \Sigma \left(f_i't_{ij}' - f_j't_{ji}' \right) \tag{8}$$

But it has already been established that some $f_i't_{ij}' - f_i't_{ji}'$ are non-zero. That means other pairs must also be non-zero to compensate. In other words, members of the ensemble must leave some states by one path and return by other paths, which constitutes a cycle.

The above line of reasoning led Morowitz to an important theorem:

In steady state systems, the flow of energy through the system from a source to a sink will lead to at least one cycle in the system.

The formation of steady state cycles has the important thermodynamic consequence that, despite large fluxes of materials and chemical transformations in the system, the net change in entropy of the *system* is zero, because entropy is a state function (as mentioned in Chapter 2), a return to the same state will always entail no net entropy change. Of course, the compensatory change in entropy of the surroundings could be greater than zero, but entropy does not *accumulate* in the system, provided that the cycles are perfect in the sense that *exactly* the same state is reproduced, which is definitely not the case in real life (see Chapter 11).

More mathematical treatments of the consequences of energy flow may be found in the writings of Ilya Prigogine and his colleagues in the Brussels School[5,6], who show how 'dissipative structures' can arise in systems far from thermodynamic equilibrium based on the analysis of entropy production. Theoretical physicist Hermann Haken[8], taking a different

44

approach, identifies 'order parameters' as macroscopic descriptors of cooperative phenomena in systems far from equilibrium, which include the Bénard convection cells we have just seen, as well as lasers (Chapter 7). In such instances, random energy fed into the system will nevertheless lead to macroscopic organization as the system passes from the thermodynamic regime of random microscopic motion of many degrees of freedom to a *dynamic* regime which has only one or a few degrees of freedom. This is also the characteristic to be expected of a system in which all the processes are linked, or *coupled* together in a symmetrical way[9]. Let us explore further what that means.

Coupled Cycles and the Steady State

Physical chemist, Lars Onsager, used the same principle of microscopic reversibility (p.41) to derive another significant result in nonequilibrium thermodynamics - the *Onsager reciprocity relation* - which shows how symmetrical coupling of processes can arise naturally in a system under energy flow[10].

The principle of microscopic reversibility can be expressed as follows: under equilibrium conditions, any molecular process and its reverse will be taking place on the average at the same rate. As we have seen in the previous Section, that implies equilibrium is never maintained by a cyclic process *at the molecular level*. To return to the three species system (p.41), the principle of microscopic reversibility, as conventionally interpreted by chemists, requires that the balance be maintained at every step:

However, the interconversions could be legitimately considered as occurring in a forward sequence, A → B → C → A, linked or coupled to a backward sequence, C → B → A → C, so that they balance each other on the whole, ie,

It turns out that the rate equations (which are like those on p.41, and I shall not write down again here) have the same form, whether one assumes detailed balance at every step or merely overall balance. In other words, *detailed molecular balance at every step is not required for thermodynamic equilibrium*. (This is what I had in mind when I drew your attention to the subtlety of the concept of thermodynamic equilibrium in the previous chapter.) So a system in equilibrium may nevertheless have within it *balanced flows* of energy and material, as long as the flows are linearly proportional to the forces. This is not surprising in retrospect, as the laws of thermodynamics were themselves a generalization of macroscopic phenomena, and do not depend on the knowledge of what is happening in detail at the molecular level.

What this means is that many *locally* nonequilibrium situations involving cyclic flow processes can be described as near approximations to the equilibrium situation, more specifically, as *fluctuations* occurring within a larger, encompassing equilibrium system, so long as the flows and forces which tend to restore the equilibrium are linearly proportional to one another.

We have come across such situations already. In the example on heat flow in a volume of an ideal gas (p. 37), a steady state is arrived at in which the convectional flow of hot molecules to the right is balanced by the diffusional movement of cooler molecules back to the left. So the transport of energy from left to right is linked or coupled to the diffusion of molecules from right to left (to maintain a concentration gradient). In the case of heat flow in a pan of water (p. 38), we have two different 'steady states' depending on the temperature difference, only the first of which - at small temperature differences - can be regarded as an approximation to equilibrium: the flow of heat by conduction from the bottom to the top is coupled to the maintenance of a density gradient in the reverse direction. Many other non-equilibrium processes may be approximated in this way. The flow of electrons across a biological membrane, for example, can couple linearly to the diffusion of inorganic ions.

Onsager set out to describe by means of general 'thermodynamic equations of motion', the rates of processes such as energy flow and diffusion, which are assumed to be linearly proportional to the 'thermodynamic forces' generating them. These forces are just the gradients of temperature, or of chemical/electrical potentials, which may be seen to be the causes of the flows. If we let J_1 and J_2 represent two coupled flow processes, and X_1, X_2, the corresponding forces, then,

$$J_1 = L_{11}X_1 + L_{12}X_2$$

$$J_2 = L_{21}X_1 + L_{22}X_2$$

(9)

The coefficients L_{11} and L_{22} are proportionality constants relating the flows each to their respective force. L_{12} and L_{21} are the cross coefficients representing the extent to which coupling occurs: in other words, how the force of one process influences the flow of the other process. For example, the two flows could be electricity and inorganic ions, due respectively to an electric potential gradient and an ionic concentration gradient. The cross coefficients tell us the extent to which the electric potential gradient influences the diffusion of ions, and conversely, how the ionic concentration gradient affects the flow of electricity.

Onsager then shows that if the principle of microscopic reversibility is true for a system in equilibrium and near equilibrium, it implies that,

$$L_{12} = L_{21}$$

This means that the coupling of the two processes becomes completely symmetric, even in states of non-equilibrium at the molecular level. In other words, the force of each process has the same *reciprocal* effect on the other process. This result can be generalized to a system of many coupled processes described by a set of linear equations of the same form as eq. (9),

$$J_i = \Sigma_k L_{ik}X_k \qquad (10)$$

where J_i is the flow of the ith process (i = 1, 2, 3.....n), X_k is the kth thermodynamic force (k = 1, 2, 3,.....n), and L_{ik} are the proportionality coefficients (where i = k) and coupling coefficients (where i ≠ k). It may happen that some of the coupling coefficients are zero (in cases where the processes do not interact at all). Some may be positive while others are negative. Still, for such a multicomponent system, certain of the couplings will be symmetrical; in other words, for *some* i and k, (i ≠ k),

$$L_{ik} = L_{ki}$$

The mathematical entities of the Onsager's thermodynamic equations of motion (eq. 10) can all be experimentally measured and verified, although it has not yet been systematically applied to the living system. Nevertheless, as we shall see, it captures a characteristic property of living systems: the reciprocal coupling of many energetically efficient processes. This raises once again the interesting question as to whether at least some of the processes in the living system may be operating near to equilibrium.

As we have seen in the previous chapter, the intricate space-time differentiation of the organism specifically allows for the establishment of a hierarchy of local near-equilibrium regimes, even though the organism as a whole is a system far from equilibrium. Perhaps one of the *raisons d'etre* of development is to set up the nested hierarchies of space-time domains where local equilibrium can be maintained in a macroscopically *non*-equilibrium system. Thus, paying attention to space-time structure leads us to a much more subtle view of both equilibrium and non-equilibrium. A system in equilibrium can have local non-equilibrium regimes; conversely, a system in non-equilibrium can also possess domains in local equilibrium.

Another intriguing question is what sort of couplings of processes can arise far from thermodynamic equilibrium? In the system of Bénard convection cells, for example, the flow of heat is obviously coupled to the convectional movement of molecules, although the flows and forces are not linearly related in such systems far from equilibrium, which include the Beloussov-

Zhabotinsky reaction (Chapter 3). Nevertheless, these 'dissipative structures' arise in such a regular, predictable way that one cannot help wondering whether a general, or 'canonical' mathematical relationship concerning the coupling of processes far from equilibrium is hovering in the wings, awaiting discovery by some mathematically able physical chemists[11].

The Manyfold Coupled Cycles of Life

Based on the above insights, we can *begin* to understand two main aspects of the living system: the ubiquitous cycling or structuring at every level of living organization, and the coupling of all these processes. This is so from the ecological cycle of the biosphere to the biochemical metabolic cycles in organisms down to the cyclic molecular machines, all meticulously choreographed, like the molecules in the Bénard convection cells, to spin and turn at different rates, each in step with the whole.

The basic outline of the ecological cycle is very simple. The photons absorbed by green plants split water molecules and reduce carbon dioxide, resulting in the formation of carbohydrates and oxygen. In respiration, the converse takes place: carbohydrates are oxidized to restore carbon dioxide and water:

$$H_2O + CO_2 \underset{\longleftarrow}{\overset{h\nu}{\longrightarrow}} \text{carbohydrates} + O_2$$

(The letters $h\nu$ represent a photon, or more accurately the energy of a photon; h, Planck's constant, equal to 6.6256×10^{-34} J s, is the smallest quantum of action and has the unit of energy in Joules multiplied by time in seconds; ν is the frequency of vibrations per second associated with the photon.)

Many secondary and tertiary cycles and epicycles feed off, or are coupled to the primary cycle above, constituting *metabolism* in living systems. Metabolism refers to the totality of chemical reactions which make and break molecules, whereby the manifold energy transformations of living systems are accomplished. The secret of living metabolism - which has as yet no equal in the best physicochemical systems that scientists can now design - is that the energy yielding reactions are always coupled to energy dissipating reactions.

The coupling can be so perfect that the efficiency of energy transfer is close to 100%. Central to the coupling of energy yielding and energy dissipating processes is the cyclic interconversion of adenosine triphosphate (ATP) and adenosine diphosphate (ADP). The terminal phosphate of ATP is added on to ADP by energy yielding processes such as photosynthesis and respiration. In photosynthesis, the energy of sunlight goes to excite electrons. As the electrons flow 'downhill' via a chain of electron transport proteins back to the ground state, the energy is tapped at several places along the way to make ATP from ADP. In respiration, similar processes of oxidative phosphorylation of ADP to ATP take place by using energy from the oxidation of complex foodstuffs. ATP is in turn converted into ADP in the biosynthesis of all the constituents of organisms and in all the energy transducing processes that enable them to grow and develop, to sense, to feel, to move, to think, to love, ... in short, to live (see Fig. 4. 7). I leave it as an exercise for readers, if they so wish, to trace out the many metabolic cycles by a careful scrutiny of a metabolic chart of 'biochemical pathways' (an assigment normally set for intransigent students).

Coupled cycles are the ultimate wisdom of nature. They go on at all levels from the ecological down to the molecular through a wide range of characteristic timescales from millennia to split seconds. Thus, the transformation of light energy into chemical energy by green plants yields food for other organisms whose growth and subsequent decay provide nutrients in the soil on which green plants depend. The energy in foodstuffs is transformed into the mechanical, osmotic, electrical and biosynthetic work both within the plants themselves and in other organisms in all the trophic levels dependent on green plants. Each kind of energy transduction in individual organisms is carried out by its own particular troupe of busy cyclic molecular machines. And upon all of these turn the innumerable life cycles of multitudinous species that make up the geological cycles of the earth.

One is reminded here of the Earth Spirit's speech in the opening scene of Goethe's *Faust*,

In the torrents of life,
in action's storm
I weave and wave
in endless motion
cradle and grave
a timeless ocean
ceaselessly weaving
the tissue of living
constantly changing
blending, arranging
the humming loom of Time I ply
and weave the web of Divinity.[12]

Energy Storage in the Biosphere

The cycling of material and energy within the biosphere automatically involves the storage, not only of material in the biosphere as biomass, but also of energy. At the beginning of this chapter, we have alluded to the large amount of energy stored in living systems, but exactly how much energy is stored? In particular, if the living system is far away from thermodynamic equilibrium, it must be free energy which is stored, for free energy is minimum at equilibrium.

There have been several attempts to estimate the relative free energy content of the biosphere. One method is to compare the energy level associated with the known molecular contents of a cell with that of the appropriate mixture of small molecular weight precursors by looking up standard tables of thermodynamic constants which have been published for many compounds. For example, Morowitz's calculations led to an estimate that the free energy level of biomass is 5.23kcal/gm higher than the starting materials. This is approximately the energetic content of one gram of carbohydrates. A slice of bread, for example, will have about 10 to 15 times this amount of energy. The *calorie* is a unit of heat energy and is equivalent to 4.184 Joules. The same kind of estimates gives values of entropy content in biomass as 0.421cal/gm *lower* than the starting materials, so one could say that biomass has a 'negative entropy' of 0.421 cal/gm. We shall examine negative entropy in greater detail in the next chapter.

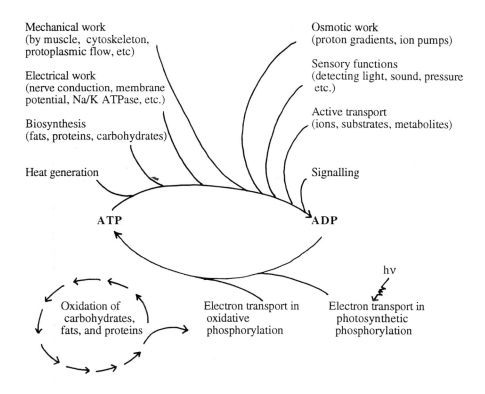

Figure 4.7 The coupled cycles of life, or energy flow in living systems.

There have also been direct measurements carried out in a 'constant volume adiabatic bomb calorimeter'. This device enables us to measure the heat of combustion of substances. It consists of a tightly sealed, heavy-walled stainless steel container, thermally isolated from its surroundings. The 'biomass', usually in a dehydrated form, is placed inside the container, which is filled with O_2 at a pressure of 30 atmospheres. The combustion is started with an electrical spark through a pair of wires in contact with the biomass. The heat released is then measured by a thermometer. From the first law of thermodynamics, the change in internal energy is equal to the heat change plus the work done:

$$\Delta U = Q + W$$

However, as the volume is constant, the work done is zero, and the decrease of internal energy is just equal to the heat released:

$$\Delta U = Q$$

These measurements give values not too dissimilar to those obtained by calculation, and represent an energy enrichment on average of 0.07eV/atom. (The electron volt, eV, is a unit of electric energy, especially useful in characterizing electronic energy levels of atoms and molecules, it is 1.602×10^{-19}J, a very minute amount of energy.)

A lot is missing from these estimates of free energy and negative entropy content. They do not take into account the space-time organization of the molecules into tissues, cells and organelles. This organization corresponds in some respect, to what Morowitz[13] has referred to as the 'configurational improbability' of the living state, in addition to its 'energetic improbability'. But far more significantly, a lot of energy is actually *stored* in the organization of the living system. For instance, gradients and potentials involving chemical species, ions and especially protons - can readily be converted into work as fluxes and flows across membrane-bound compartments to drive the transport of metabolites or movements of cilia and flagella. Nevertheless, let us take the estimate as a reasonable figure for the moment, and see what it tells us about energy flow on earth.

If we designate the average energy enrichment of biomass as ε, and the average residence time of the energy in the system as t, then the necessary flow rate per atom is,

$$f = \varepsilon/t \qquad (11)$$

This has the same form as the chemical equation for the flow of species of atoms and molecules through a system,

$$\text{Flow of species} = \frac{\text{Total amount of the species in the system}}{\text{mean residence time}} \qquad (12)$$

Unlike chemical species, however, energy cannot be 'tagged', for example, with a radioactive label, and its fate followed through the system; so the residence time for energy cannot be measured directly. However, as the flow of energy into the biosphere is always accompanied by the flow of materials - especially CO_2 - into the system, the mean residence time for energy can be taken as the mean residence time for carbon in the system. The size of the various carbon pools on the surface of the earth has been estimated, giving the total biomass (both living and dead) on land and in the sea as 2.9×10^{18} gm and 10.03×10^{18} gm respectively. The values for carbon flow, i.e., the total fixed by photosynthesis per year, on land and in the ocean, are respectively, $0.073 \pm 0.018 \times 10^{18}$ gm and $0.43 \pm 0.3 \times 10^{18}$ gm. Putting these values into eq. (12) above gives residence times of 40 years and 21.8 years. In terms of energy, an average flow of 0.003 eV per atom of biomass per year suffices to maintain the level of the biomass. This flow corresponds to a total annual flow of energy of 4.8×10^{17} kcal, a value in good agreement with the estimated yearly fixation of energy by photosynthesis, which is $13.6 \pm 8.1 \times 10^{17}$ kcal[14].

An interesting question arises here: what is the significance of the long residence time of the energy that comes to the biosphere in photons from the sun? The energy of the photon meanders through innumerable cycles and epicycles of metabolism such that it is released and stored in small packets (in the ATP molecules, for example) ready for immediate utilization or in longer term depots such as gradients and fields to yet longer term deposits in the form of glycogen and fat. The efficiency (and perhaps stability) of metabolism is associated with this drawn-out web of coupled energy transfer, storage and utilization within the highly differentiated space-time structure of the organism. Metabolic and structural complexity prolongs the energy residence or storage time, perhaps by an equal occupation of *all* storage times, affording the organism an efficient and stable living. We can see an interesting connection here back to the concept of stored energy versus thermalized energy introduced in Chapter 3 (p. 27). *As long as the energy remains stored, it can be utilized for work.* We shall have the occasion to return yet again to the importance of energy storage for vital functions in later chapters.

A similar significance may well attach to the ecological cycles where the stability of the planetary system is concerned. There has been a great deal written on ecological sustainability, as well as the need to preserve genetic diversity within the past several years. Diversity may be much more important for the homeostasis - and hence sustainability - of planet earth than is generally recognized. The residence time of the energy within the biosphere is directly related to the energy stored, and hence, to species diversity or equivalently, the size of the trophic web (see eq. (12)), which, on the planetary level, is the space-time organization of the global ecological community. Could it be that in the 'geophysiology' of planet earth there is the same wisdom of the body that is in metabolism? The relationship between complexity of ecological communities and stability has already captured the attention of some theoretical ecologists[15], though few attempts have yet been made to take space-time organization into account.

In the next chapter, we shall examine more closely the molecular basis of energy and entropy, where it will become clear that the quantity of energy in biomass as such cannot be the whole story to the success of living organisms. It is the quality of the energy, the structure of the living system and the way energy is stored and mobilized in flows and fluxes that are most crucial for life.

Notes

1. See Morowitz (1968).
2. Lovelock (1979).
3. Fox (1986).

4. The account given is a simplified version of that given in Morowitz (1968) pp23-25.

5. See Prigogine (1962).

6. See Nicolis and Prigogine (1989) p. 12.

7. The following account is based on that given in Morowitz (1968) pp.29-33.

8. Haken (1977).

9. I am indebted to Kenneth Denbigh for suggesting that coherence in the living system may have something to do with the coupling of processes. This encouraged me to study Onsager's reciprocity relation more carefully and started me thinking about the consequences of coupling in equilibrium and non-equilibrium systems.

10. Onsager (1945). See Denbigh (1951) for a very clear and accessible account on which the following description is based.

11. Geoffrey Sewell has recently obtained a non-linear generalization of the Onsager reciprocity relations for a class of irreversible processes in continuum mechanics. See Sewell (1991).

11. MacDonald (1989).

12. See Morowitz (1978) pp. 252-3.

13. See Morowitz (1968) pp. 68-70.

14. See May (1973).

CHAPTER FIVE

HOW TO CATCH A FALLING ELECTRON

Life and Negative Entropy

In the last chapter we saw how energy flow leads to material cycling and energy storage in living systems. But in what forms is the energy stored, and how is energy storage related to living organization?

Here is Schrödinger's famous statement about life:
"It is by avoiding the rapid decay into the inert state of 'equilibrium' that an organism appears so enigmatic....What an organism feeds upon is negative entropy. Or, to put it less paradoxically, the essential thing in metabolism is that the organism succeeds in freeing itself from all the entropy it cannot help producing while alive."[1]

In a footnote Schrödinger explains that by 'negative entropy', he really means *free energy*. "But," he continues,
"this highly technical term seemed linguistically too near to *energy* for making the average reader alive to the contrast between the two things. He is likely to take free as more or less an *epitheton ornans* without much relevance, while actually the concept is a rather intricate one, whose relation to Boltzmann's order-disorder principle is less easy to trace than for entropy and 'entropy taken with a negative sign'..."[2]

Despite Schrödinger's apology for the term 'negative entropy', it continues to be used, and by the most authoritative among scientists:
"It is common knowledge that the ultimate source of all our energy and negative entropy is the radiation of the sun. When a photon interacts with a material particle on our globe it lifts one electron from an electron pair to a higher level. This excited state as a rule has but a short lifetime and the

electron drops back within 10^{-7} to 10^{-8} seconds to the ground state giving off its excess energy in one way or another. Life has learned to catch the electron in the excited state, uncouple it from its partner and let it drop back to the ground state through its biological machinery utilizing its excess energy for life processes."[3] So writes Nobel laureate biochemist Albert Szent-Györgi, who has inspired more serious students in biochemistry than any other single person.

What exactly is this negative entropy, and how is it related to free energy and the manner in which the living system avoids the decay to equilibrium? Schrödinger uses the term 'negative entropy' in order to describe a somewhat fuzzy mental picture of the living system, which not only seems to avoid the effects of entropy production - as dictated by the second law - but to do just the opposite, to increase organization, which intuitively, seems like the converse of entropy. Szent-Györgi, on the other hand, has conveniently included the notions both of free energy and of organization in his use of the term. I think that both scientists have the right intuition - energy and organization are inextricably bound up with each other. Hence energy is not something associated merely with molecules. In order to appreciate that, we need to follow through the argument of those who have attempted to translate thermodynamic concepts, such as energy and entropy, into molecular motions and configurations.

Many thermodynamicists have already cautioned us (see Chapter 2) that the thermodynamic entropy, $\Delta S = Q_{rev}/T$, for example, has no direct connection with the statistical mechanics term by the same name, $S = k \ln W$[4], nor with notions of 'order' and 'disorder', much less with 'organization'.

Actually, even the relationships among all the thermodynamic entities: entropy, free energy, total energy, heat and temperature, are by no means straightforward. The physicist-philosopher, P.W. Bridgman, has often been criticized for pushing the 'operationalist' positivist approach in science, which says that science ought only to be about entities that one can measure or define by means of an operation. However, Bridgman's primary motivation was to expose the shaky foundations of fuzzy concepts to which scientists attribute an undeserved and misplaced concreteness, extrapolating the concepts indiscriminately to situations where they may no longer apply.

Scientists, he says, should always keep before themselves the 'man-made' nature of their science (instead of treating it as though God-given). He has presented a thorough and rigorous critique of the thermodynamic concepts in his book, *The Nature of Thermodynamics*[5], and as far as I am aware, no one has successfully replied to his criticisms[6]. Let us now take a closer look at the thermodynamic concepts and their statistical mechanical interpretations.

Free Energy and Entropy

In Chapter 2, I have written down some equations for the change in free energy which are most relevant for biological systems: the Helmholtz free energy for processes occurring at constant temperature and volume,

$$\Delta A_{sys} = \Delta U_{sys} - T\Delta S_{sys} \tag{1}$$

and the Gibbs free energy for processes at constant temperature and pressure,

$$\Delta G_{sys} = \Delta H_{sys} - T\Delta S_{sys} \tag{2}$$

The criterion for all spontaneous processes is that free energy always decreases, ie,

$$\Delta A_{sys} < 0$$

and,

$$\Delta G_{sys} < 0$$

Or, in the limiting case that the process occurs at equilibrium, the free energy change is zero. Processes which involve a negative free energy change (loss of free energy to the surroundings) are said to be *exergonic* - and are thermodynamically 'downhill' or spontaneous; those which involve a positive free energy change, or gain in free energy, are *endergonic*, and are thermodynamically 'uphill' and non-spontaneous.

As the free energy term is a difference between two other terms: total energy or enthalpy, and entropy, we have to see how the two change in a reaction. For example, if ΔH is negative and $T\Delta S$ positive, they reinforce each other and ΔG will also be negative, so the reaction would proceed spontaneously. In other situations, ΔH and $T\Delta S$ may work against each other, then ΔG would depend on the relative magnitudes of the two terms. If $|\Delta H|$ >> $|T\Delta S|$, then the reaction is said to be enthalpy driven because the sign of ΔG is predominantly determined by ΔH. Conversely, if $|T\Delta S|$ >> $|\Delta H|$, then the reaction is entropy driven. In case of an adiabatic process occurring reversibly, which involves no heat exchange with its surroundings, $Q = 0$, hence $\Delta S = 0$, and the change in free energy is simply equal to the change in internal energy or enthalpy, ie,

$$\Delta A = \Delta U$$

$$\Delta G = \Delta H$$

We have already mentioned such reactions in the living system towards the end of Chapter 3.

There are, in fact, no absolute measures for the quantities such as U, H, and G which enter into the equations of thermodynamics. Only the *changes* can be calculated and a zero point for the quantity is arbitrarily fixed at standard conditions of 298.15K and one atmosphere. The one exception may be entropy.

According to Boltzmann's equation given in Chapter 2,

$$S = k \ln W \tag{3}$$

entropy is proportional to the logarithm of the number of microstates, W, in the macroscopic system. In a perfect crystal, there can be only one arrangement of the atoms and so there must only be one single microstate at absolute zero, ie,

$$S = k \ln 1 = 0$$

This gives rise to the so-called third law of thermodynamics, which states that:

Every substance has a finite positive entropy, but at the absolute zero of temperature, the entropy may become zero, and does so become in the case of a perfect crystalline substance[7].

Not all substances have only one microstate at absolute zero. Nevertheless, one can appreciate that the entropy of a substance is related to the random thermal motion of the molecules - a thermal energy that is somehow not available for work, and tends to disappear at absolute zero temperature. This is in opposition to the free energy, which *is* somehow available for work. But as pointed out above, there need be no entropy generated in an adiabatic process. From that we suspect that the division into available and nonavailable energy cannot be absolute: the energy associated with a molecule simply cannot be partitioned into the two categories.

The second law identifies the direction of processes which can occur spontaneously. However, it says nothing about the *rate* of the processes. Many thermodynamically favourable, or downhill processes do not actually proceed appreciably by themselves. The reason is that every substance sits inside its own individual energy well, separated from others by a hump which is the *energy barrier* (see Fig. 5. 1). In order for a reaction to take place, an amount of energy, known as the *activation energy*, must first be supplied to overcome this barrier.

This is why the rate of almost all reactions goes up with temperature. With increase in temperature, the random motion of the molecules becomes exaggerated, increasing the likelihood of collisions on which reaction depends. This is the basis of Arrhenius' law of the dependence of reaction rates on temperature. But if increasing the temperature increases reaction rates, then not all random thermal motion can be regarded as entropy, some of that must actually contribute to the *free energy of activation*, and indeed, such a term has been invented for chemical kinetics. So again, this warns us that useful energy and the random kinetic energy that goes to make up entropy cannot be so neatly distinguished.

Why Don't Organisms Live by Eating Diamonds?

This question was raised by F. Simon, who pointed out to Schrödinger that his simple thermodynamical considerations cannot account for our having to feed on matter in the extremely well-ordered state of more or less complicated organic compounds rather than on charcoal or diamond pulp, which are crystalline substances with regular molecular order and consequently, have little or no entropy[8].

After all, according to thermodynamic principles, organisms could make their living by eating carbon or diamond pulp (for negative entropy), and absorbing energy as heat directly from its surroundings, say, by taking up

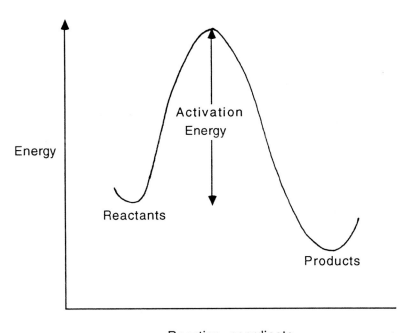

Figure 5.1 Activation energy for chemical reactions.

residence inside a volcano. So there may be something special about the precise form, or quality, of energy that is taken up, or how that energy is taken up by living systems that perhaps enables them to build organization, *which is not the same as crystalline order.* For one thing, the former is dynamic through and through, whereas the latter is static.

To explore further the interrelationships among the concepts of heat, energy, entropy, negentropy and organization, we need to know something of the description as to how energy is distributed at the molecular level.

According to *the principle of equipartition of energy,* which is based on classical (as opposed to quantum) theory, the energy of a molecule is equally divided among all types of motions or *degrees of freedom.* For mon-atomic gases, each atom has three translational degrees of freedom. For molecules containing more than one atom, there will be other motions, notably, rotation (moment) about the centre of mass on the interatomic bonds and vibration (stretching and compressing) along the bonds. For a molecule containing N atoms, we need $3N$ coordinates to describe all its complete motion. Of these, three are for translational motion, three angles are needed to define the orientation of the molecule about the three mutually perpendicular axes through the centre of mass, and that leaves $3N - 6$ degrees of freedom for vibrations. If the molecule is linear, only two angles are needed to specify rotation, leaving $3N - 5$ degrees of freedom for vibrations.

For one single molecule, each translational and rotational degree of freedom possesses energy $\frac{1}{2}kT$, whereas each vibrational degree of freedom possesses energy kT. This is so because the vibrational energy contains two terms, one kinetic and one potential, each of which is $\frac{1}{2}kT$. For one mol. of a gas (the molecular weight in grams), there are N_0 molecules, where N_0 is Avogadro's number and is equal to 6.02217×10^{23}; and the corresponding energies for each degree of translation and rotation is $\frac{1}{2}N_0kT$, or $\frac{1}{2}RT$, and for vibration, RT, where $R = N_0k$ is the gas constant and is equal to 8.314 J $K^{-1}mol^{-1}$.

We can now calculate the total internal energy U for any system of gases. For a diatomic gas such as O_2, for example,

$$U = \tfrac{3}{2}RT + RT + RT = \tfrac{7}{2}RT$$

(translation) (rotation) (vibration)

How can we test whether the derivation of U based on the kinetic theory of gases corresponds to the same entity in thermodynamics? This can be done by heat capacity measurements. The specific heat of a substance is the energy required to raise the temperature of 1g of the substance by 1 degree. For the chemist, the *mole* is a more convenient unit of mass, and the corresponding specific heat is the molar heat capacity, C, the energy required to raise 1 mol of the substance by 1 degree. The heat capacity at constant volume is defined by the equation,

$$C_v = (\partial U/ \partial T)_v$$

where $\partial U/ \partial T$ represents the partial derivative of internal energy with respect to temperature.

According to the derivation above, for a diatomic gas,

$$C_v = (\partial U/ \partial T)_v = \tfrac{7}{2}R = 29.10J\ K^{-1}\ mol^{-1}$$

It turns out that the predicted values agree with experimental measurements for monatomic gases, but show considerable discrepancies for molecules containing two or more atoms (see Table 5. 1).

The discrepancies can be explained on the basis of quantum theory, which we shall deal with in more detail in a later chapter. For the moment, it is sufficient to recognize that according to quantum theory, the electronic, vibrational and rotational energies of a molecule are 'quantized', which means that they do not exist in a continuum, but only at discrete levels (see Fig. 5.2). Thus, the spacing between successive electronic levels is much larger than that between vibrational energy levels; which is in turn larger than that between rotational energy levels. The spacing between successive translational energy levels is so small that the levels practically merge into a continuum. In that respect, translational energy is classical rather than quantum mechanical.

When a system absorbs heat from its surroundings, the energy is used to promote various kinds of molecular motion. The heat capacity is hence an energy capacity, or the system's capacity to store energy. Theoretically, energy may be stored in any mode. But it is much easier to excite a molecule to a higher rotational energy level than to a higher vibrational or electronic energy level.

Table 5. 1   Calculated and measured heat capacities of gases at 298K[9]

Gas	C_v (J K^{-1} mol^{-1}) Calculated	C_v (J K^{-1} mol^{-1}) Measured
He	12.47	12.47
Ne	12.47	12.47
Ar	12.47	12.47
H_2	29.10	20.50
N_2	29.10	20.50
O_2	29.10	21.05
CO_2	54.06	28.82
H_2O	49.87	25.23
SO_2	49.87	31.51

Hence, molecular energies can at least be quite unambiguously classified in terms of the *levels of excitation and storage*. Quantitatively, the ratio of the occupation numbers (ie the number of molecules in each energy level), N_2/N_1, in any two energy levels, E_2 and E_1 is given by Boltzmann's distribution law:

$$N_2/N_1, = e^{-\Delta E/kT}$$

where $\Delta E = E_2 - E_1$, k is Boltzmann's constant and T the absolute temperature. For translational motion, ΔE is about 10^{-37}J, so $\Delta E/kT$ at 298K is,

$$\frac{10^{-37}\text{J}}{(1.38 \times 10^{-23}\text{JK}^{-1})(298\text{K})} = 2.4 \times 10^{-17}$$

The resulting figure is so close to zero that the value for the ratio, N_2/N_1, is very nearly unity. In other words, the successive levels of translational energy are about equally populated at room temperatures. On the other hand, for

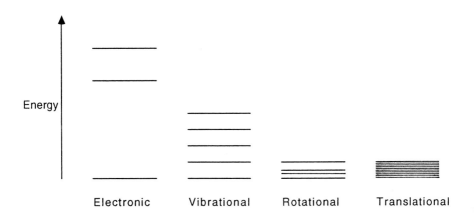

Figure 5.2 The quantized energy levels for the different forms of molecular motion[10]. The lowest energy levels, associated with translational motion, are so close together that they form a continuum. Translational energies are hence, classical.

vibrational energy, where ΔE is about 10^{-20}J, repeating the calculation done above gives $\Delta E/kT = 2.431$, so N_2/N_1 is much smaller than unity in this case. That means only a very few molecules will be in the higher energy levels at 298K. For electronic energies, the population of higher energy levels is even much less likely, and at 298K, almost all of the molecules will be in the lowest energy levels.

From these considerations, one would expect that at room temperatures, we can neglect both vibrational and electronic energies, so the energy of the diatomic system becomes,

$$U = \tfrac{3}{2}RT + RT = \tfrac{5}{2}RT$$
$$\text{(translation)} \quad \text{(rotation)}$$

This give $C_v = \tfrac{5}{2}RT = 20.79$J K^{-1} mol^{-1}, which is quite close to the measured value (see Table 5.1).

When the temperature increases, however, the vibrational levels will be expected to make a substantial contribution to the heat capacity. And this is indeed the case. At 1500K, the heat capacity of O_2 closely approaches the theoretical value of 29.10 JK^{-1} mol^{-1}. At 2000K, the measured value of 29.47 JK^{-1} mol^{-1} actually exceeds the theoretical value, suggesting that electronic motion is beginning to make a contribution to heat capacity (see Table 5.2).

The alert reader will have already noticed by now that the above account describes quite adequately what happens when energy is supplied as heat to *a system of unorganized molecules*, which is what physics and chemistry usually deal with. The heat absorbed saturates all the lower energy levels before the upper levels become populated, and a large spectrum exists ranging from the translational energies at the low end to the electronic energies at the high end. The higher the level at which energy is stored, the less the increase in temperature per unit of energy stored, ie, the greater the heat capacity. Much of the energy stored in living system is in the high electronic levels. In that respect alone, in the language of quantum theory, the living system has achieved a 'population inversion' contrary to Boltzmann's law, ie, the upper energy levels are much more populated than they should be for the temperature of the system, which is typically about 300K.

Table 5. 2.   Measured heat capacities of O_2 at different temperatures[11]

T (K)	C_v (J K^{-1} mol^{-1})
298	21.05
600	23.78
800	25.43
1000	26.56
1500	28.25
2000	29.47

It has been said that our bodies have a high 'electronic' temperature and that if we were equilibrium thermodynamic systems, we would have a temperature of up to 3000K. But that too, is misleading. For at those temperatures, the high molecular weight constituents could no longer exist; they would be subject to denaturation and dissociation long before those temperatures are attained. And perforce neither cells nor tissues could exist.

With regard to the wide spectrum of energies associated with the molecules, one finds that a priori, they cannot be clearly apportioned into free energies, or entropy, or heat; much less can we say which parts correspond to work. These can only be apportioned a posteriori - after the reactions have taken place, and furthermore, depending on how the reactions have taken place. That may seem contrary to the textbook assertion that unlike work or heat, entropy is a state function; which means that it is a property of the state as much as its temperature, volume and pressure, and does not depend on how that state is arrived at. However, there are no operational measures of absolute entropy content, except by spectroscopic methods at absolute zero. And only changes in entropy are defined in terms of a reversible process. Yet,

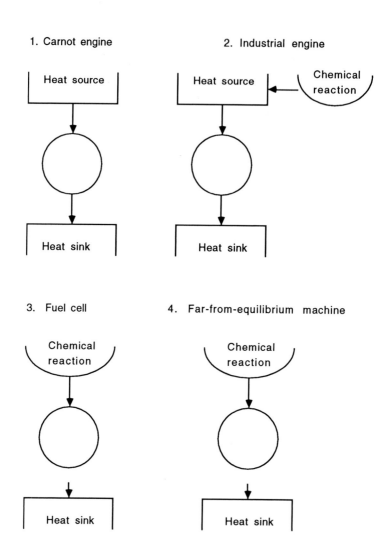

Figure 5.3 Four types of engines. Carnot and industrial engines are subject to Carnot's efficiency theorem as they depend on heat exchange. The fuel cell and far from equilibrium machines do not depend on the conversion of energy into heat, and hence are not subject to the same constraints (see text). The incomplete arrows leading from the fuel cell and far-from-equilibrium engines to the heat sink indicate that the heat loss is not a necessary part of the working cycle[12].

there are many natural irreversible processes in which the change in entropy simply cannot be defined (see Chapter 11). So, the change in entropy is just due to the energy which has gone missing, or has somehow become dissipated. To work out how much has gone where is rather like an accountant balancing the books at the end of the year! And just as it is misleading to base next year's budget on this year's accounts, one cannot hope to explain how the organism works by laws governing steam engines, or an unorganized collection of molecules. Let us try to be more precise about this.

What Kind of 'Engine' is the Organism?

In discussing the limitations of the second law of thermodynamics as usually formulated, Morowitz lists 4 types of engines (see Fig. 5.3). The first three, the Carnot engine, the industrial engine and the fuel cell, are all equilibrium devices. As the first two engines operate by converting chemical energy into heat which is then converted into work, they are both subject to *Carnot's efficiency theorem*, which places an upperbound to the efficiency - when operated reversibly - as $(1 - T_{sink}/T_{source})$. In other words, efficiency is determined by the temperature difference between source and sink, or the boiler and the exhaust in the heat engine.

The fuel cell, however, operates isothermally, converting chemical energy directly into electricity, and so its efficiency is no longer limited by the temperature difference between source and sink. Instead, the maximum efficiency is given by $(1 - T\Delta S/\Delta U)$, where ΔS and ΔU are the changes in internal entropy and energy of the fuel cell and T is the temperature of the surroundings.

The fourth type is the far-from-equilibrium machine that can in principle operate at very high efficiencies without any heat loss. Living systems, Morowitz suggests, could be isothermal equilibrium machines, like the fuel cell; or it could be a non-equilibrium, type 4 machine. "It is also possible," he continues, "that they operate on a molecular quantum mechanical domain that is not describable by any of the macroscopic engines..."[13]

Energy transfer via heat, on which the science of thermodynamics is based, is by far the least efficient and nonspecific form of transfer. The biosphere, as we have seen, does not make its living by absorbing heat from the environment. No organism can live like a heat engine, nor can it obtain its energy or negative entropy by feeding on carbon or diamond pulp and burning it with oxygen. Instead, life depends on catching an excited electron quite precisely - by means of specific light absorbing pigments - and then tapping off its energy as it falls back towards the ground state. Life uses the highest grade of energy, the packet or quantum size of which is sufficient to cause specific motion of electrons in the outer orbitals of molecules. It is on account of this that living systems can populate their high energy levels without heating up the body excessively, and hence contribute to what Schrödinger intuitively identifies as 'negative entropy'. But what enables living systems to do so? It is none other than their meticulous space-time organization in which energy is *stored* in a range of time scales and spatial extents. This is what I mean at the beginning of the previous chapter (p. 34) when I say that energy flow organizes the system which in turn organizes the energy flow. Stored energy is in the organization, which is what enables the living system to work so efficiently on a range of timescales.

It is of interest to compare the thermodynamic concept of 'free energy' with the concept of 'stored energy'. The former cannot be defined *a priori*, much less can be it be assigned to single molecules, as even *changes* in free energy cannot be defined unless we know how far the reaction is from equilibrium. 'Stored energy', originally defined by McClare with respect to a characteristic time interval (see Chapter 3), can readily be extended, in addition, to a characteristic spatial domain. As such, stored energy is explicitly dependent on the *space-time structure of the system*, hence it is a precise concept which can be defined on the space and time domain of the processes involved. Indeed, stored energy has meaning with respect to single molecules in processes involving quantum molecular machines (see p. 30) as much as it has with respect to the whole organism. For example, energy storage as bond vibrations or as strain energy in protein molecules occurs within a spatial extent of 10^{-9} to 10^{-8}m and a characteristic timescale of 10^{-9} to 10^{-8}s. Whereas in terms of a whole organism such as a human being, the overall energy storage domain is in metre-decades.

In the previous chapter, we also discovered how the living system, though macroscopically far from thermodynamic equilibrium, may yet harbour a hierarchy of local domains of equilibrium defined by the space-time magnitude of the processes involved. Thus, equilibrium and non-equilibrium may be less important a distinction between living and non-living systems, than space-time structure, or organization. Perhaps what we really need, in order to understand the living system, is a *thermodynamics of organized complexity* which describes processes in terms of *stored* energy instead of *free* energy. (Alternatively, we can define the hierarchy of local equilibria in terms of the energy storage domains.) Thus, the living system could contain isothermal equilibrium machines, non-equilibrium machines, as well as quantum molecular machines (see p.30). We shall explore these ideas further later on.

The crucial difference between living and non-living systems lies in how energies are stored, channeled and directed. It is of interest to note that Lord Kelvin, one of the inventors of the second law, has stated it as follows: "It is impossible by means of an inanimate material agency to derive mechanical effect from any portion of matter by cooling it below the temperature of the coldest of the surrounding objects." Lest there should be any ambiguity as to what is intended by the reference to 'inanimate material agency', he adds, "The animal body does not act as a *thermodynamic engine*...whatever the nature of these means [whereby mechanical effects are produced in living organisms], consciousness teaches every individual that they are, to some extent, subject to the direction of his will. It appears therefore that animated creatures have the power of immediately applying to certain moving particles of matter within their bodies, forces by which the motions of these particles are directed to produce derived mechanical effects."[14] The secret of the organism is that it can mobilize the whole spectrum of energies for work, from the translational to the electronic, making scant distinction between 'bound' and 'free' energy, between that which is 'unavailable' and otherwise.

To both summarize and anticipate the story thus far: 'negative entropy' in the sense of organization, or simply 'entropy with the sign reversed' has not so much to do with free energy as Schrödinger indicates, but with the way energy is trapped, stored and mobilized in the living system. Energy is trapped directly at the electronic level. It is stored not only as electronic bond energies, but also in the structure of the system: in gradients, fields and flow patterns, compartments, organelles, cells and tissues. All this in turn enables organisms to mobilize their energies *coherently* and hence make available the entire spectrum of stored energies for work, whenever and wherever energy is called for.

Notes

1. Schrödinger (1944) pp.70-71.

2. Schrödinger (1944) p. 74.

3. Szent-Györgi (1961).

4. K. Denbigh is of this opinion, although Oliver Penrose (personal communication) disagrees! Both scientists are authorities in the field of thermodynamics and statistical mechanics. This illustrates precisely the complexity of the issues involved.

5. Bridgman (1961).

6. Oliver Penrose disagrees with this statement (personal communication). In his book, *Foundations of Statistical Mechanics,* he has shown that a non-decreasing function corresponding to entropy can be derived from a Gibbs ensemble by the mathematical technique of 'coarse-graining'. See also Penrose, (1981).

7. This statement is due to Chang (1990) p. 140.

8. See Schrödinger (1944) p. 74.

9. From Chang (1990) p. 43.

10. Modified from Chang (1990) p. 44.

11. From Chang (1990) p. 45.

12. Modified from Morowitz (1978) pp. 75.

13. See Morowitz (1978) pp. 74-6.

14. Cited in Ehrenberg (1967) p. 104.

CHAPTER SIX

THE SEVENTY-THREE OCTAVES OF NATURE'S MUSIC

Reduction versus Integration

Western science is dominated by an analytical tradition of separating and fragmenting that which for many of us, appears to be the seamless perfection that once was reality[1]. Yet beneath this reductionistic tendency runs a strong countercurrent towards unity and integration, particularly within the past hundred or so years. I am motivated to write this book partly by a confluence of ideas which is drawing together the hitherto divergent streams of physics, chemistry and biology. I have a strong feeling that this is how we can begin to understand the organism, and in so doing, restore to some extent the sense of unity that once gave meaning to life.

The two great unifying concepts in physics and chemistry that are most relevant for understanding the organism are the laws of thermodynamics and the electromagnetic theory of light and matter. We have dealt with thermodynamics at some length in the preceding chapters, showing, among other things, how the organism differs from a thermodynamic engine and from a collection of unorganized molecules. In this and the chapters following, we shall concentrate on the electromagnetic theory of light and matter, to show how electromagnetic energy animates and coordinates the living system. Electromagnetic theory also places both light and matter within the central conundrum of the wave particle duality of physical reality. This has far reaching implications on the nature of biological processes, indeed, of consciousness and free will. Before we launch into examining the relevance of electromagnetic energy for living systems and the deep philosophical issues raised by the quantum theory of reality, let us review classical electromagnetic theory as put forward by James Clerk Maxwell.

Electricity and Magnetism

The fundamental relationship between electricity and magnetism was discovered in the last century from observations on *electromagnetic induction*. Charges moving in a wire induce a magnetic field around the wire. Conversely, moving a conducting wire through a magnetic field induces a current to flow in the wire. These phenomena are brought together first by Faraday, and then formalized in Maxwell's electromagnetic theory. Maxwell reasons that moving charges in the wire give rise to an electric field surrounding the wire, which induces a magnetic field. The magnetic field in turn, induces another electric field which induces a further magnetic field, and so on. Actually, the continuous mutual induction only occurs in an oscillating electrical circuit, or from a pair of oscillating charges. This is because, just as the charges have to be moving in order to induce a magnetic field, the magnetic field has to be changing to induce an electric field. Figure 6.1 gives the electric field lines in the neighbourhood of a pair of oscillating electrical charges. The spreading electric field lines are accompanied by magnetic field lines which are at right angles to them. This gives rise to electromagnetic waves that propagate away from the oscillating charges.

In the course of this work, Maxwell made a second sensational discovery concerning light. The phenomenon of light has been studied from the time of the Greeks, and after many experiments, two competing theories were advanced. One maintains that light consists of tiny, invisible particles that move along rays, like a continuous series of minute shots fired from a gun. The other theory claims that light is a motion of waves propagating from a source. Both theories explain *reflection*, as from a mirror, and *refraction*, the change in the direction of light rays when passing from one medium to another, say from air to water. However, the *diffraction* of light - light bending around an obstacle - is more reasonably explained by a wave theory. But if it were a wave, there must be a medium through which it propagates through space. This medium was given the name of *ether*. From astronomical studies, the velocity of light had been determined in 1676 by the Danish astronomer Roemer to be about 186,000 miles per second. Maxwell found that the velocity of electromagnetic waves is also 186,000 miles per second. The identity of these figures and the fact that both electromagnetic

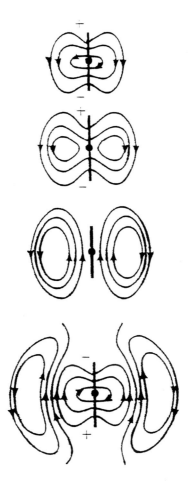

Figure 6.1 Electric field lines near a pair of oscillating charges. The lines have the same appearance in any plane containing the charges. The lines of the magnetic field induced are perpendicular to those of the electric field. The electric and magnetic fields propagate together as waves through space. The sequence is read from top to bottom.

radiation and light were known to be wave motions immediately suggested to Maxwell that light must consist in the transverse undulations of the same medium (ether) which is the cause of electric and magnetic phenomena. In other words, light consists of a succession of electric and magnetic fields. With one stroke of genius, Maxwell had connected phenomena as widely disparate as cosmic radiation, sunlight, the microwave oven, radio and television broadcast, and all the uses of electricity. The electromagnetic spectrum starts from wavelengths of 10^{-14}m at one extreme to 10^8m at the other, spanning a range of 10^{22}. In terms of doublings, $10^{22} \approx 2^{73}$, or 73 octaves. This is the range of nature's music, of which we ourselves are part, and which we can tune into in its entirety, for better or for worse. The organism's own music - of a somewhat more restricted but still enormous range between 10^{-7}m and 10^8m and beyond - is both exquisite and subtle. With the present level of 'man-made' electromagnetic pollution of the environment, one might very well ask whether the organism's melody is in grave danger of being drowned out altogether.

The Electromagnetic Theory of Light and Matter

The 50 years straddling the end of the last century and the beginning of the present century must have been a heady era for western science. It began with Maxwell's theory that light is a form of electromagnetic radiation, essentially the same as that which is emitted by an oscillating electric circuit, only at much higher frequencies. This wave theory of light was soon thrown into doubt, however, by its failure to account for a number of observations, among which, is the so-called blackbody radiation emitted by all bodies at temperatures above absolute zero, and the photoelectric effect, the ejection of electrons from a metal surface when light of a definite threshold frequency is shone upon it (for the theory of which Einstein was awarded the Nobel prize). Both these phenomena require the explanation offered by Planck: that the energy in electromagnetic radiation comes only in packets, or *quanta*, the size of which depends on the frequency, i.e.,

$$E = h\nu$$

where E is energy, h is Planck's constant, and v, the frequency. These quanta are absorbed, one by one, as though they were particles, resulting in the ejection of an equal number of electrons in the photoelectric effect:

$$h v = W + \tfrac{1}{2}mv^2$$

where W represents the energy that the photon must possess in order to remove an electron from the metal and $\tfrac{1}{2}mv^2$ is the kinetic energy of the ejected electron, m being its mass, and v its velocity. W is a measure of how strongly the electrons are held in the metal. Thus, one must regard light not only as waves, but also as particles under some circumstances.

This dual character - the wave particle duality of light - was soon extended to matter. Particles of matter, under certain circumstances, will also exhibit wave properties. De Broglie proposed in 1924 the following relations:

$$\lambda = h/mv = h/p$$

where p is the momentum, mass x velocity, λ is the wave length, and v the frequency. This was experimentally confirmed several years later when diffraction patterns of electron waves by aluminium powder were obtained. The electron microscope, now routinely used in biology, is based on just this wave-like behaviour of accelerated electrons.

In parallel with these developments, the structure of the atom was also being defined. It was known in the 1930s that atoms consist of a nucleus containing positively charged particles, called *protons*, and neutral particles, called *neutrons*. As atoms are electrically neutral, there must be an equal number of negatively charged particles, or electrons present. These are found outside the nucleus. In order to account for the emission spectra of excited atoms, which were known since the 19th Century, and for the stability of atoms, Niels Bohr incorporated Planck's hypothesis to develop his theory of the atom. Our present picture is essentially a later refinement of the Bohr atom, described in 1913.

Essentially, Bohr hypothesized that electrons are not found just anywhere outside the nucleus, but are confined to 'shells' for which the *angular momentum* is given by

$$p = mvr = nh/2\pi \tag{1}$$

where m and v are the mass and velocity of the electron, r is the radius of the shell, n is an integer, 1,2,3.... Within these shells, the electrical attraction between the electron and the nucleus just balances the outward acceleration due to the circular motion of the electron, ie,

$$Ze^2/r^2 = mv^2/r \tag{2}$$

where Z is the atomic number (the number of protons in the nucleus), e is the electronic charge. Combining equations (1) and (2) gives the permitted radii of successive shells, ie,

$$r_n = \frac{n^2h^2}{4p^2me^2Z} \tag{3}$$

Thus, the radii of stable shells are proportional to n^2, ie, $1^2, 2^2, 3^2, ...$ or 1, 4, 9... whereas the angular momentum, p, goes up in proportion as n, ie, 1, 2, 3... (see eq. (1)). The ground state is $n = 1$. It is closest to the nucleus and corresponds to a radius of $r = 0.529Å$, which is called *Bohr's radius*. Successively higher energy levels are represented by $n = 2, 3, 4, ..$ and so on. The total energy of the electron at each level is the sum of the kinetic and potential energies, ie,

$$E = \tfrac{1}{2}mv^2 - Ze^2/r \tag{4}$$

The negative sign associated with the potential energy (the second term on the right of the above equation) indicates an attractive interaction between the electron and the nucleus. From equation (2), we have,

$$mv^2 = Ze^2/r \tag{5}$$

Substituting into equation (4) gives,

$$E = -Ze^2/2r = \tfrac{1}{2}mv^2 \tag{6}$$

Combining equations (3) and (6) gives,

$$E_n = \frac{-\ 2mp^2Z^2e^4}{n^2h^2} \tag{7}$$

Thus, the energy levels go up in proportion as -1, -1/4. -1/9 and so on. At sufficiently large distances away, the energy is almost zero, and that is where the electron, being no longer attracted to the nucleus, becomes mobile. In other words, it becomes a little electric current until it drops into a vacant electron shell of another atom. Usually, only electrons in the outermost shell can be excited into mobility. The difference between the energy level at infinity and the outermost populated electron shell defines the *ionizing potential* of the atom (see Fig. 6.2).

The shells are successively filled with electrons as we proceed along the periodic table of the elements in accordance with four quantum numbers: n, l, m_l and m_s. The value of n gives the total energy of the electron; l, its angular momentum, m_l determines the angle between the angular momentum of the electron and an external magnetic field, and hence the extent of the magnetic contribution to the total energy of the atom. All three numbers are quantized. Thus, n can take any integer from 1 to n, l can take any integer from 0 to n-1, and m_l can take any integer from -1 through 0 to +1. In addition, each electron carries a spin, described by the fourth quantum number, m_s, which can take values of $+\tfrac{1}{2}$ or $-\tfrac{1}{2}$, according to the two opposite directions of spin. The other major principle determining the distribution of electrons around an atom is *Pauli's exclusion principle*: no two electrons in an atom can exist in the same quantum state. This means that each electron

80

in a complex atom must have a different set of the four quantum numbers. The quantum number n defines the energy level of a shell, within which different subshells are specified by the quantum number l, and within each subshell, m_l defines pairs of states, of opposite spins, or $m_s \pm \frac{1}{2}$. In this way, the atomic structure of all the elements in the periodic table can be specified. The first shell ($n=1$) has only one subshell containing the single electron of hydrogen. Adding another electron of opposite spin gives helium, which completes the first shell. The next element, lithium, has three electrons, so the additional one has to go to the $l = 0$ subshell of the $n = 2$ shell, which has two subshells, the other being $l = 1$, and so on.

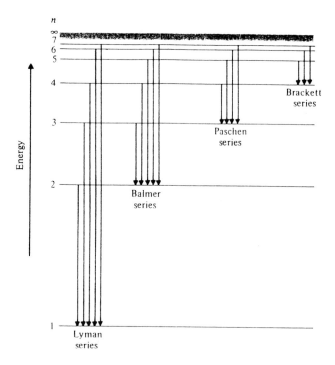

Figure 6.2 Energy levels of the electron shells in a hydrogen atom, and the different emission spectra obtained when excited electrons move from a higher to a lower energy shell.

A final word of caution about the picture of the atom presented above: it would be wrong to think of the electron as a minute charged particle orbiting the nucleus, possessing definite position and momentum. Quantum theory tells us, instead, it is more like a complex probability cloud distributed over the whole of the three-dimensional orbital, as consistent with the wave-particle duality of all matter (as well as light). We shall deal with the wave particle duality in more detail in Chapter 10.

Molecular and Intermolecular Forces

Atoms are most stable when their electron shells are closed or fully filled, and tend to gain or lose electrons in order to secure closed shells by joining with other atoms. This is the basis of chemical reactivity. Two principal binding mechanisms are known, *ionic* and *covalent*. In ionic bonds, electrons are effectively transferred from one atom to another, eg,

$$Na + Cl = Na^+ + Cl^-$$

The resulting positive and negative ions attract each other with an electrostatic force given by Coulomb's law:

$$F = Q_1 Q_2 / er^2$$

where F is the force; Q_1, Q_2 are the electric charges of the ions; r is the distance between them and e is the dielectric constant (or, to physicists, the permittivity) of the medium. The dielectric constant measures how effectively the medium shields the negative and positive charges from one another and hence reduces the force between them. Water, a major constituent of living systems, is an excellent dielectric because of its large effective *dipole moment* (see below). Water molecules can surround the ions in solution and effectively neutralize their charges.

In a covalent bond, which is the majority of chemical bonds in biological molecules, two atoms share one or more pairs of electrons in order to provide each with a closed outer electron shell. Molecules formed by covalent bonds, in which the centres of the positive and negative charges are separated, or do not exactly coincide, are said to be dipolar, and possess a *dipole moment* :

$$m = Q \times r$$

where Q is the charge separated and r the distance between them. Dipolar molecules interact electrostatically with one another, the magnitude of the force being proportional to $1/d^4$, where d is the distance separating the two molecules. The magnitude of the dipole moment in each case depends on the difference in electronegativity between the atoms involved in bonding. The order of electronegativity is, for example,

$$H < C < N < O$$

Water has a dipole moment of 1.85 debye (where 1 debye = 3.38 x 10^{-30} Coulomb meter). Polypeptide chains in the α-helical configuration have enormous dipole moments upwards of 500 debye, as the individual moments of the peptide bonds are all aligned. In the double helical DNA, on the other hand, the antiparallel arrangement of the two strands means that there is no net dipole moment, even though the single strands have their dipole moments due to the sugar-phosphate bonds in the backbone all aligned in the same direction[2] (see Fig.6.3).

A third bonding mechanism is the hydrogen bond, which arises from dipole interactions in molecules where hydrogen is bonded to an electronegative atom such as oxygen, or nitrogen. This results in the bonding electrons being closer to the other atom than to the hydrogen, leaving a net positive charge on the non-bonded side of the hydrogen atom. The latter can thus be involved in bonding with other electronegative atoms, constituting the hydrogen bond. Hydrogen bonds form between water molecules, giving rise to supramolecular aggregates (see Fig. 6.4).

Hydrogen bonds are responsible for stabilizing the α-helical secondary structure of the polypeptide chain, as well as the β-pleated sheet structures between polypeptide chains. They also stabilize various 'conformations' or folded tertiary and quarternary structures of polypeptides and proteins. The hydrogen bonds between base-pairs in the DNA double helix are responsible for the precise templating mechanism which ensures the faithful reproduction of the base sequence of the DNA molecule during replication.

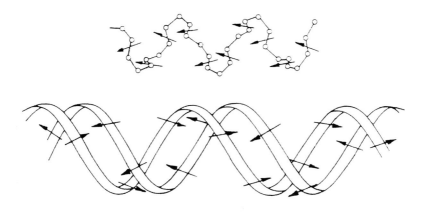

Figure 6.3   The dipole moments of the α-helix in a polypeptide (above) and a double helix of the DNA (below)[3].

Finally, nonpolar molecules, or molecules which are electrically neutral, are nonetheless subject to short range attraction by so-called *dispersion* or *London forces*. It turns out that these forces can be explained in terms of the instantaneous configuration of the field around the atom, creating temporary dipoles. One especially important form of bonding due to London forces in the living system is the *hydrophobic* (literally, water fearing) interaction, which accounts for why oil and water do not mix. Hydrophobic interaction is responsible for structuring lipid molecules into biological membranes, and for the folding of polypeptide chains into specific shapes or conformations so suited to their functioning.

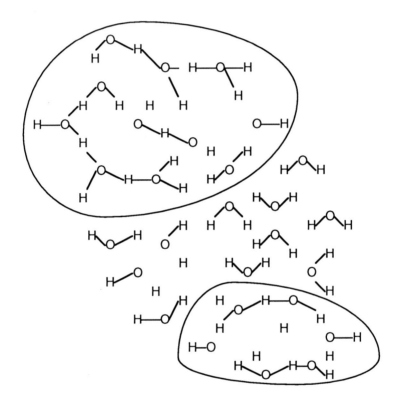

Figure 6.4 Supramolecular aggregates of water molecules formed by hydrogen bonding[4].

Thus, electromagnetic forces underlie all molecular and intermolecular interactions. The situation is best summarized in 1966 by the chemist, Peter J.W. Debye, after whom the unit of dipole moment is named[5]:

"Forty years ago discussions started about the interpretation of van der Waals' universal molecular attraction as a result of the Coulombic interactions of the electrical components of the molecule. The average electrical field around a molecule was analyzed, which led to its characterization by dipole-, quadrupole- and higher moments. The mutual orientation effect of such molecules was recognized as a reason for attraction. However, the higher the temperature, the smaller the effect of such an orientation has to become. So the conclusion to be drawn was that at high enough temperatures molecular attraction should vanish. This obviously was in contradiction to the experimental facts, and so the picture of molecules as rigid electrical structures was abandoned. Instead, their structure was recognized as deformable under the influence of an electric field: polarizability was introduced. The result was that, under the combined influence of the electrical field carried by one molecule and the polarization induced by it in its partner an additional potential energy emerges which is proportional to the square of the field and drives the molecules to each other. This still was not enough, since it made the mutual attraction of single atoms like neon or argon, around which the average field is zero, not understandable. The last difficulty was overcome by London when he recognized that the field important for the interaction should be taken as the instantaneous field. This introduced, unavoidably, quantum theory as a new and essential background...

"It was at this juncture that colloid chemistry entered into the field with the recognition that the attraction between colloid particles was the result of essentially the same molecular forces as those which determine, for instance, the heat of vaporization of a liquid. Calculations appeared concerning the van der Waals attraction between spheres and plates, giving rise to a series of brilliant experiments, which directly measured van der Waals' forces between plates at distances of the order of the wavelength of visible light, and which also demonstrated that indeed van der Waals' attraction is a result of the electromagnetic interaction of the molecular stray fields.

"However, the theory of molecular interaction had first to be refined. The molecular field introduced by London was an electrostatic field. Casimir recognized that its finite velocity of propagation had to be taken account of.

86

One of the most characteristic results of this refinement is the calculation of van der Waals' attraction between two perfect mirrors. This attraction depends solely on two fundamental constants, Planck's quantum of action h and the light velocity c, and illustrates emphatically how intimate the relation is between van der Waals' universal attraction and the quantum fluctuations of the electromagnetic field.

"In recent times observations on the critical opalescence have also entered the field as appropriate for the measurement of molecular forces. In the vicinity of the critical point [for phase transition], interactions are observed which are apparently of long-range character extending over distances of the order of the wavelength of visible light. ...this can be understood as a result of long-range correlation, based on short-range interactions..."

The last paragraph in Debye's summary makes it clear that the same electromagnetic attractive forces are responsible for long range order during equilibrium phase transitions. We shall continue our exploration of how these same intermolecular forces may be involved in the organization of the living system in the next chapter.

Notes

1. I have written on this myself, see Ho (1992a).
2. see Pethig (1979).
3. From Pethig (1979) p. 50.
4. Modified from Chang (1990) p. 500
5. Debye 's Foreword to Chu (1967).

CHAPTER SEVEN

THE COHERENT EXCITATION OF THE BODY ELECTRIC

Collective versus Statistical Behaviour

The living system is a bewildering mass of organized heterogeneity. Its molecular diversity alone would defy description in terms of any statistical mechanisms involving large numbers of identical species. Within the confines of a volume less than 1 μm^3 bound by a cell wall of woven complex carbohydrate fibres, each *E. coli* bacterium has a single copy of a gigantic DNA molecule consisting of 10^6 non-periodic base-pairs (monomeric units) coding for the 10^3 different cellular proteins, the majority of which exists in no more than 100 copies, and a minority considerably less. Within the same volume, there are 10^3 species of RNAs, mostly represented by a few copies; and in addition, there are diverse membrane lipids, carbohydrates, fats and other metabolites, small molecular weight cofactors, and inorganic ions, all together in a minute, highly indented and wormholed droplet of water.

And yet, this vastly complicated mixture of molecules, within a volume smaller than a pin-head, behaves with such order and regularity that Schrödinger concludes it must be "guided by a 'mechanism' entirely different from the 'probability mechanism' of physics."[1] What is the mechanism that Schrödinger has in mind? It is none other than the collective behaviour of physical systems at low temperatures when molecular disorder - entropy - disappears and the systems no longer behave statistically but in accordance with dynamical laws. These collective behaviours are the subject matter of solid state physics, or condensed matter physics, as it is called nowadays.

Thus, the molecules in most physical matter have a high degree of uncoordinated, or random thermal motion; but when the temperature is lowered to beyond a critical level, all the molecules may condense into a collective state, and exhibit the unusual properties of *superfluidity*, when all the molecules of the system move as one, and *superconductivity*, in which electricity is conducted with zero resistance (see Fig. 7.1) by a perfectly coordinated arrangement of conducting electrons. Liquid helium, at temperatures close to absolute zero, is the first and only superfluid substance known thus far; and various pure metals and alloys are superconducting at liquid helium temperatures. Today, technology has progressed to superconducting materials which can work at $125K^2$.

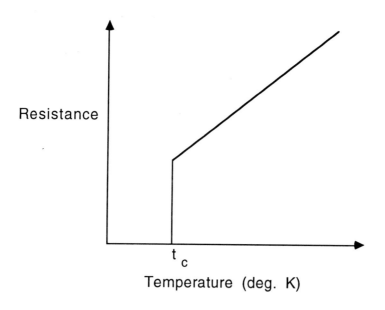

Figure 7.1 Graph of electrical resistance versus temperature in a superconductor.

Since Schrödinger's time, a number of people have suggested that collective modes of activity may arise in living systems which are analogous to those occurring at low temperatures in physical systems. But how can that take place, given that living systems typically work at around 300K? A clue is supplied by the observation that nonequilibrium systems subject to energy flow can also undergo transitions to coherent, macroscopic activity, as we have seen in the Bénard convection cells in Chapter 4.

Non-equilibrium Transitions to Dynamic Order

The similarity between equilibrium and nonequilibrium phase transitions has been noticed by Hermann Haken[3], as both involve an abrupt change from a state of molecular chaos to one with macroscopic order - as though the different subsystems are cooperating with one another. Equilibrium phase transitions include the familiar solid-liquid, liquid-solid changes of states all pure substances as well as mixtures of substances undergo at characteristic temperatures; other less familiar examples are transitions to ferromagnetism (see p. 117), superfluidity and superconductivity at low enough temperatures. All of these involve a reduction in the number of possible microstates in the system ultimately to a single one. The type of order achieved is essentially static - being that of a perfect crystal. In the nonequilibrium system, however, the transition is always to a regime of *dynamic* order where the whole system is in coherent motion. Let us consider laser action, which is representative of non-equilibrium phase transitions.

In a solid state laser, specific atoms are embedded in a block of solid-state material (Fig. 7.2), at the ends of which are reflecting mirrors. Energy can be supplied in the form of light, electric current, or even heat, in order to excite the atoms. The atoms re-emit light tracks and those running in the axial direction will be reflected back and forth several times before going out. At low levels of energy pumping, the laser operates as an ordinary lamp, and the atoms emit randomly. As the pumping power is increased, a level is reached, called the laser threshold, when all the atoms will oscillate in phase and emit together, sending out a giant wave train that is 10^6 times as long as that emitted by individual atoms. In this example, as in the Bénard convection cells described in Chapter 4, we can clearly see the transition from a regime where individual molecular motions are random with respect to one another, and hence no macroscopic or long range order is present, to a regime where all the molecules are moving coherently, with the concomitant dramatic appearance of long range macroscopic order. In both systems, random energy is channelled into coherent modes of activity at phase transition. This is a characteristic feature of *non*equilibrium phase transitions as it makes them stable to thermal noise or other perturbations, thus distinguishing them from equilibrium phase transitions such as superconductivity and ferromagnetism which are destroyed by thermal energies. The two nonequilibrium systems, however, differ in an important respect. The Bénard convection cells result from a *classical* phase transition, which typically occurs over a long timescale. The solid state laser, by contrast, results from a *quantum* phase transition which takes place very rapidly. This difference in timescale, in my opinion, tips the balance in favour of biological organization being essentially a quantum rather than a classical phenomenon (see later).

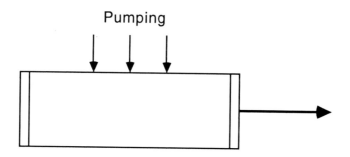

Figure 7.2 A solid state laser.

Can it be that similar non-equilibrium phase transitions to macroscopic dynamic order occur in living systems? And if so, what may be the sort of energy involved? The solid-state physicist Herbert Fröhlich[4,5] points out that, as living organisms are made up predominantly of dielectric, or dipolar molecules packed rather densely together, they may indeed represent special solid state systems where electric and viscoelastic forces constantly interact. Under these conditions, metabolic pumping may result in condensation to collective modes of vibration, rather like the laser action just described.

Specifically, thermal energies arising from metabolism can be retained in the system by the excitation of giant dipolar molecules such as proteins, nucleic acids and cellular membranes, which typically have an enormous electrical field of some 10^7V/m across them. The excited molecules vibrate at various characteristic frequencies involving the coupling of electrical displacements and mechanical deformations. This eventually builds up into collective modes of both electromechanical oscillations (phonons, or sound waves) and electromagnetic radiations (photons) that extend over macroscopic distances within the organism and perhaps also outside the organism. The radiated mode arises because oscillating charges always emit electromagnetic waves, as we have seen in the last Chapter. Fröhlich refers to these collective modes as 'coherent excitations'. There are three kinds of collective modes. The first is a stable or metastable highly polarized state where the separation between positive and negative charges are maximum; this results from mode softening of interacting frequencies towards a collective frequency of 0. The second is a *limit cycle* oscillation, a limit cycle being a cycle which has a stable orbit that neither gets smaller nor bigger. The third mode arises when the energy supply exceeds a certain threshold, when oscillations of much higher frequencies occur. Each collective 'mode' can be a band of frequencies, with varying spatial extents, as consistent with the spatiotemporal structure of the living system. Nevertheless, the frequencies are coupled together so that random energy fed into any specific frequency can be communicated to other frequencies. This is reminiscent of non-equilibrium phase transitions in which random energy becomes channelled into macroscopic order as in the Bénard convection cells and the solid state laser described above, though conceptually, Fröhlich achieves energy exchange via a 'heat-bath', a quasi-equilibrium approximation in thermodynamics (see Chapters 2 and 4).

What we must imagine is an incredible hive of activity at every level of magnification in the organism - of music being made using perhaps two-thirds of the 73 octaves of the electromagnetic spectrum - locally appearing as though completely chaotic, and yet perfectly coordinated as a whole. This exquisite music is played in endless variations subject to our changes of mood and physiology, each organism and species with its own repertoire. It would be wonderful if we could tune in and discover how some of us are made of Schubert, and others of Beethoven, or Bach.....

Coherent excitations can account for many of the most characteristic properties of living organisms that I have drawn your attention to at the beginning of this book: long range order and coordination; rapid and efficient energy transfer, as well as extreme sensitivity to specific signals. An intuitive way to appreciate how coherence affects the rapidity and efficiency of energy transfer is to think of a boat race, where it is paramount for the oarsmen and oarswomen to row in phase; and similarly, in a line of construction workers moving a load, as the one passes the load, the next has to be in readiness to receive it, and so a certain phase relationship in their side-to-side movement has to be maintained. If the rhythm (phase relationship) is broken in either case, much of the energy would be lost and the work would be slower and less efficient. Songs and dances may have their origin primarily in the pleasurable sensations arising from the entrainment of one's own internal rhythms to those of the collective, they also serve to mobilize the energies of the collective in an efficient and coherent way.

How would coherent excitations make the system sensitive to specific, weak signals? Such a weak signal will be received by the system only when the system is 'in tune' - rather like a very sensitive radio receiver, which can resonate to the signal. Furthermore, a small signal will be greatly amplified for it will not only affect one molecule, but because many other molecules are in the same state of readiness, they too, will be affected in the same way, and the signal is correspondingly multiplied as many times as there are molecules responding to it. A situation where this is most dramatically illustrated is the crystallization of a solid from a supersaturated or supercooled solution: a tiny speck of dust would nucleate the process all at once in the entire volume of fluid. Something analogous may be involved in the functioning of living systems, except that there would be much more specificity. Whole populations of cells may be poised in critical states so that a small, specific signal would set off a whole train of macroscopic, coherent reactions.

The idea of coherence is so foreign to most western-trained scientists that there is a lot of resistance to it from the mainstream. It is easy to blame this just on the reductionistic tendencies of western science for which cooperativity and coherence are anathemas. There is a more immediate aspect that comes from the practice and methodology of biological sciences itself, which is all of a piece with the conceptual framework.

'The Cataclysmic Violence of Homogenization'

As biochemists brought up in the early 1960s, we were schooled to the idea that the cell is, to all intents and purposes, a bag of concentrated solution of enzymes and metabolites mixed up at random, save for a few organelles and intracellular membranes. Enzymes, by the way, are proteins that catalyse all the chemical reactions in the body; each enzyme being generally specific for a single reaction. Enzymes enable reactants to overcome often very large activation energy barriers so that the reaction can take place at physiological temperatures. In other words, enzymes are the main reason why organisms can work isothermally, at physiological temperatures (and pressures). On account of their important role in metabolism, much of biochemistry is given over to the study of the mechanism of enzyme action *in vitro*, (that is, isolated and outside of the cell), although how they actually work *in vivo*, within the cell, is still a mystery. As good biochemists then, we found this 'mixed-upness' - which we supposed to exist inside the cell - an irritating obstacle to analysis and is something one has to overcome. The standard procedure is to grind up the organisms or cells to a pulp, or 'homogenate' - in which everything is well and truly mixed up, and then proceed to separate out the different 'fractions' according to size or density, to extract, and to purify until we have stripped off all contaminating 'impurities' surrounding the enzyme of interest, to end up triumphantly with a single species 'purified to homogeneity'.

This precious preparation is then dissolved in pure, double-distilled, deionized water to which ultrapure substrates are added, and the long process of 'characterization' of enzyme activity begins, based on the notion of random diffusion of enzyme and substrate molecules through the aqueous solution. All this of course, goes to reinforce the idea we begin with - that the cell is a bag of enzymes and metabolites dissolved in solution.

As electron microscopy and other specific staining techniques became available, it gradually dawned on us that the cell is highly structured. Nowadays, the generally accepted picture of a cell is quite sophisticated. It is bound by the cell membrane - a double layer of lipids, which is supported by and attached to the membrane skeleton composed of a basketwork of contractile filamentous proteins lying immediately underneath it. The membrane skeleton in turn connects with the three-dimensional network of various fibrous proteins collectively known as the *cytoskeleton*, which links up the inside of the cell like a system of telegraph wires terminating onto the membrane of the nucleus. In the nucleus, the chromosomes (organized complexes of DNA and proteins) are anchored directly to the inside of the nuclear membrane. The nuclear membrane and the cell membrane are also in communication via concentric stacks of membranous vesicles, the *Golgi* apparatus - with special secretory functions, and the *endoplasmic reticulum* - a system of three-dimensional canals and spaces believed to be involved in intracellular transport and occupying a large proportion of the cell volume. A substantial volume is also taken up by organelles such as the *mitochrondria*, where simple carbohydrates are oxidized to CO_2 and H_2O with the generation of ATP, and *ribosomes* on which polypeptide chains are synthesized. Finally, what is left over is the *cytosol* (or 'soluble' cytoplasm). Despite this much improved picture, most biochemists still believe that organelles are made up of structural proteins which are relatively inert, and that the 'functional' proteins or enzymes are dissolved at random in the cytosol, where the main metabolic theatre of the cell is being played out. There are a few possible exceptions, of course, such as the proteins involved in oxidative and photosynthetic phosphorylation, which are found inside the mitochondria and the chloroplasts respectively.

However, a number of different lines of investigation began to suggest that perhaps no protein is dispersed at random in solution, but practically all of them are organized in specific multiprotein complexes, and furthermore, a high proportion of the cell water may actually be bound or structured on the enormous amount of surfaces within the cell[6].

Intermolecular complexes giving rise to the submicroscopic cellular organization are usually attributed to steric factors specific to the different molecular species involved. And textbooks in biochemistry are filled with beautiful diagrams to illustrate this static, 'lock and key' principle of 'recognition' between enzymes and their respective subtrates, between signal molecules and their respective receptor proteins, as well as between different species of proteins. Yet, everything in quantum theory tells us that molecules never stand still, and proteins are no exception. Indeed, the most mobile proteins are enzymes which generally have great specificities for their substrates, and the most mobile parts of the immunoglobulins or antibody molecules are precisely those containing the antigen recognition sites[7], which is very difficult to reconcile with the lock and key model.

What is more likely to be responsible for intermolecular complexes in cellular organization is the universal electromagnetic attractive forces between molecules that we have considered in the last chapter. This also provides an extremely potent source of specificity and selectivity within the electromagnetic spectrum itself. Even a small fraction of the 73 octaves would far out-perform any specificity in shapes of molecules. It is already well-known that molecules with the same intrinsic frequency of vibration not only resonate over long distances, and hence undergo coherent excitations, but can also attract one another over long distances (see Chapter 3). Furthermore, frequency specificity is generally accurate to within 1%[8]. Some biochemists studying enzyme kinetics are coming around to the idea that coherent excitations may have more relevance for enzyme action than the conventional 'lock and key' principle[9]. The importance of recognition by resonance cannot be over-emphasized. Thermal energies, particularly at equilibrium, possess no information whereas an excitation of a specific frequency at a temperature where no other excitation of the same energy exists - i.e., in a system far from equilibrium - not only has just the requisite information to do the work, but the inherent power to do it as well[10]. It provides the motive force of attraction between appropriate reactive species *and* enables energy exchange to take place. Thus, 'information' is not something separate from energy and organization.

The most general considerations would tell us that structuring, or 'self-assembly' is the most probable, one might say, inevitable consequence of the range of intermolecular forces that exists. So part of the 'negentropy' that Schrödinger was searching for is simply located in the chemistry of matter, which physicists have a tendency to forget. For example, bilayer lipid membranes form spontaneously at water/water interfaces to maximize intermolecular hydrophobic and hydrophilic interactions, with the nonpolar side chains of the lipid molecules buried within the membrane, and the polar groups exposed to the water on either side. Hydrophobic interactions between lipids and proteins, similarly, account for the localization of many proteins in membranes. Hydrogen bonding is responsible for supramolecular aggregates, as we have seen, resulting in complexes in which the dipole moments of individual bonds are either vectorially oriented in the same direction, as is the case with various α-helical contractile proteins, or they can form antiparallel complexes in which their dipole moments neutralize each other. Intramolecular and intermolecular hydrogen bonding are also involved in stabilizing protein conformations. Spontaneous self-assembly of supramolecular structures such as microtubules and actin filaments (both important constituents of the cytoskeleton) may involve such oriented dipole interactions and can indeed take place outside the living organism[11]. Similarly, the adsorption of water molecules onto the surfaces of membranes and proteins is to be expected on the basis of polar interactions, just as much as water is excluded from the interior of proteins on the basis of hydrophobic interactions between the nonpolar amino acids.

For many years now, the physiologist, James Clegg[6], among others, has championed the idea that the cell is meticulously ordered down to the detail of individual protein molecules. It is, in effect, a solid state system[12]. Within this system, each protein has its own community of other proteins and small molecular cofactors and metabolites. In other words, an enzyme in isolation - stripped of its rich 'cytosociology'[9] - is merely a shadow of its real identity. Its natural highly diverse neighbourhood is the key to its efficient functioning. This is a fascinating microcosm of the same principle of diversity that operates in ecology (see Chapter 4), and perhaps also in societies. (My own experience is that multicultural societies make for a richer social and intellectual life and are more vibrant than closed monolithic communities.)

Clegg summarizes a plethora of evidence suggesting that even the cytosol is filled with a dense network of actin-like protein strands, or "microtrabecular lattice", connecting with almost all cytoplasmic structures and to which practically all of the enzymes and many of the substrates are attached (see Fig. 7.3). The surface area of the microtrabecular lattice is estimated to be about 50 times that of the cell, and its volume about 50% that of the cytosol. Thus, very little of the macromolecular constituents of the cytosol is dissolved in the cell water. On the contrary, up to 60% of the cell water itself might be bound, or structured by the surfaces of the microtrabecular lattice, the intracellular membranes, and the cytoskeleton.

Inspite of, or perhaps because of its highly condensed and connected nature, the whole cell is extremely dynamic, the connections between the parts as well as the configurations of the cytoskeleton, the membranes, the chromosomes, and so on, can all be remodelled within minutes subject to appropriate signals from the environment, as for example, the presence of food and light, hormones, growth factors, mechanical or electrical stimulation (see below). The entire cell acts as a coherent whole so that information or disturbance to one part propagates promptly to all other parts.

The evidence for this organized description of the cytosol comes from many sources, including electronmicroscopy under mild, nondisruptive conditions, measurements of physical parameters of the cell water relative to 'free' water outside the cell, and various cytological manipulations to establish the location of cellular enzymes and substrates. I shall only describe two kinds of experiments, which seem to me the most decisive.

The first involves the centrifugation of live cells such as the unicellular alga, *Euglena*, which has a cell wall thick enough to withstand the centrifugal forces, and hence remains viable. After centrifuging at 100,000xg (where g is the earth's gravitation force) for 1 h - the standard condition used in the laboratory - the cellular constituents became stratified into several

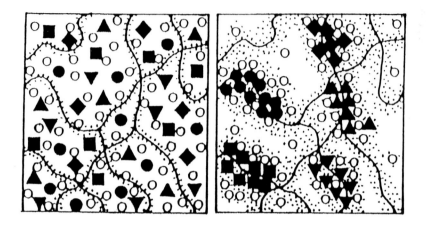

Figure 7.3 Diagrammatic representation of the soluble (left) versus
solid state (right) description of the cytosol. Wavy lines are
cytoskeletal elements of the microtrabecular lattice (see text), dots are
structured water molecules, open circles are metabolites and
cofactors, and filled symbols are macromolecules[13].

zones according to density (see Fig. 7.4), which is as expected. The unexpected
finding is that the soluble phase near the top was found to contain *no*
macromolecules at all: sensitive tests for the presence of a long list of
enzymes, proteins, and nucleic acids were all negative. This contrasts with
results from the routine fractionation of the homogenate - obtained by
pulping the cell under generally very disruptive conditions - in which the
100,000g supernatant or soluble phase contains many proteins, enzymes as
well as nucleic acids. Centrifuging other intact, live cells gave essentially
identical results: no macromolecules were found in free solution. As
remarked by one advocate of the organized, solid state cell,

"..the empirical fact that a given molecule appears primarily in the 'soluble' fraction may divert attention from the cataclysmic violence of the most gentle homogenization procedure."[14]

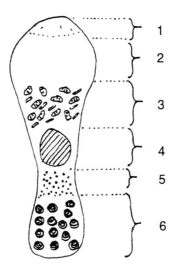

Figure 7.4 Diagram of the layers of cellular structures formed after centrifugation of *Euglena*. 1. Fat droplets, 2. Soluble supernatant, 3. Mitochondria, Golgi apparatus and endoplasmic reticulum, 4. Nucleus, 5. Ribosomes, 6. Starch grains[15].

In another kind of experiment, holes were created in the cell membrane by mild treatments with Dextran sulphate (a complex sulphated carbohydrate) in such a way that the cells remained alive, so that when the Dextran sulphate was removed, the cells recovered and resealed themselves. The holes were big enough to allow subtrates and metabolites to go through. Yet, 60 mins. after treatment, the metabolic rate of the permeabilized cells was only decreased by about 50% compared to controls, suggesting that not only are the macromolecules or enzymes bound within the cell, but many of the small molecular weight metabolic intermediates and cofactors may also be bound or compartmentalized so that they cannot diffuse freely away from the cell even when the membrane is full of holes.

The consequence of structured enzyme-substrate complexes is that intermolecular dipole interactions will be very important in determining metabolic function. In particular, a high level of correlation of movements will exist between successive enzymes in a metabolic pathway and between enzyme and substrate, making for the characteristically rapid and efficient catalysis in living organisms which has yet to be achieved in the test-tube. Also, structured water layers on membranes and macromolecular surfaces transmit dipole interactions or oscillations and proton currents could flow in the layer of water molecules immediately next to membranes or macromolecules, as suggested by observations on both oxidative and photosynthetic phosphorylation[16,17]. The flow of protons along membranes has been experimentally demonstrated in artificial monolayers made of membrane phospholipids[18] as well as along hydrated proteins and nucleic acids[19].

The evidence for the organized cytosol comes from methods of study designed to minimize disruption to endogenous cellular structure so that the cell can be observed as nearly as possible in its natural functioning state. It is a general approach towards the ideal of 'non-invasiveness', where the cell or organism is allowed to tell its own story, as it were, to inform us of its internal processes as it is living and developing. As techniques become more and more sensitive and precise, we can for the first time, analyse without destroying. The picture we obtain is far from atomistic and reductionist. On the contrary, we begin to appreciate the full extent of the cooperativity and coherence of living processes right down to the scale of individual molecules.

'Life is a Little Electric Current'

So far, we have been concentrating on electrodynamical forces. Should we not also consider other types of forces and flows such as heat or mechanical energy, diffusion or convection? We have dealt at some length in previous chapters on why heat transfer cannot be the major form of energy transduction in living systems. This does not preclude thermal energies being used for work, as in a coherent system, even thermal energies can be channelled into coherent excitations. However, heat conduction and heat convection cannot play a substantial role, as dielectric activities will come into effect long before. (Besides, living systems are, to all intents and purposes, isothermal (see Chapter 5), especially in warm-blooded animals where special mechanisms to generate heat are involved in keeping the body warm. This emphasizes that heat is not an inevitable byproduct of metabolic reactions.) For the same reason, passive diffusion is simply too slow to account for the rapidity of energy transfer. Some or all of these processes may be involved in the back-up activities to restore the pre-existing steady state, or in feed-forward activities leading to longer term changes. Mechanical deformations of dielectric molecules do occur of course, but they are always accompanied by electromagnetic effects. The main reason for concentrating on electrodynamical forces is simply that the predominant molecular and intermolecular forces are all electromagnetic, and in the living system which approaches the solid state, these interactions are expected to be by far the most important. Furthermore, energy first enters into the biosphere as light, which is electromagnetic radiation. Living systems may indeed be electrodynamical through and through. As Szent-Györgi[20] says,

"..life is driven by nothing else but electrons, by the energy given off by these electrons while cascading down from the high level to which they have been boosted up by photons. An electron going around is a little current. What drives life is thus a little electric current."

If the organism is in effect a solid state system, then many of the basic principles of condensed matter physics will have applications in the living system. The important difference we should always bear in mind, is that the living system has a much more complex and dynamic organization that somehow enpowers it to metabolize, grow, differentiate and maintain its individuality and vibrant wholeness, something no physical system can yet do. Nevertheless, let us explore what condensed matter physics can tell us about the living system.

We have given reasons as to why intermolecular dipole attractive forces may be responsible for much of the cellular organization. We also know that mobile charges, carried by both electrons and protons, play a major role in energy transduction and transformation in the living system. Szent-Györgi among others, has noted that the sites of primary energy transduction, the cell membranes, are closely analogous to the pn junction, a semiconductor device which facilitates the separation of positive and negative charges, and is capable of generating an electric current when excited by energy in the form of heat or light[21]. It is the basis of the solar cell, which should prove to be the cleanest and most efficient source of renewable energy, if that is the lesson we can obviously learn from nature.

In common with these semi-conductor devices, various biological membranes and artificially constituted phospholipid membranes also exhibit thermoelectric, photoelectric and piezoelectric effects due respectively to heat, light and mechanical pressure[22]. In addition, many semiconductor devices are luminescent, producing light as the result of heating, electrical pumping (the basis of electroluminescent devices which are used in producing laser light) and also stimulation by light[21]. We shall see later that organisms do indeed exhibit luminescence, which can be stimulated in a highly nonlinear way with heat and light. The basis for generating electricity in all cases is the electron in the outermost orbital of atoms and molecules - the valency electron - which becomes mobile on absorbing enough energy (corresponding to the ionizing potential). Mobile electrons constitute electric currents, which can flow to the nearest neighbouring molecule, or it can go further afield.

Actually, the electric currents in living organisms flow simultaneously in two opposite directions because there are both positive and negative charges. The molecule from which the electron is lost - the *electron donor* - acquires a net positive charge, or *hole*, in the language of solid state physics; while the molecule to which the electron goes - the *electron acceptor* - acquires a negative charge. Both positive and negative charges can propagate further by moving down an electronic cascade represented schematically in Figure 7.5. The molecules, D, A_1, A_2 and A_3 with decreasing electronic energy levels, sit in close juxtaposition in a biological membrane (as is the situation in the electron transport chain involved in photosynthetic and oxidative phosphorylation). The molecule D, on being excited, donates a valency electron from its ground level to the excited level of A_1. The hole in D is filled by an electron coming from another molecule on the ouside to the left, while the excited electron progresses through the excited levels of the molecules, A_2 and then A_3. Positive and negative charges propagate in opposite directions. Electric currents are thus readily generated by such mechanisms of charge separation and transfer which are quite rapid, as they involve no substantial molecular reorganization or chemical change that would otherwise have slowed the process down and caused a lot of unnecessary disturbance.

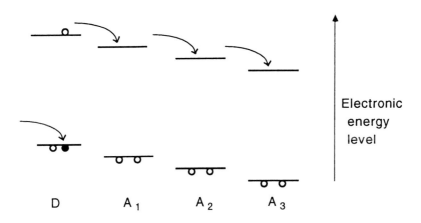

Figure 7.5 The electronic cascade. The molecule D is the electron donor, while A_1, A_2 and A_3 are the electron acceptors.

The first step in photosynthesis is just this kind of charge separation and vectorial transfer: a chlorophyll molecule, on absorbing a photon, donates an electron to an acceptor which flows down the electron transport chain, generating ATP on the way. The 'hole' left in the donor chlorophyll is filled by an electron from another donor on the outside, i.e., water. The 'hole' in this case becomes a proton, which is also mobile, and it can either accumulate in the bulk phase on one side of the membrane to create a gradient for the transport of metabolites and ions across the membrane, or it can flow in a local circuit along the surface of the membrane. Some people believe that in addition to electricity or the flow of electrons, substantial *proticity* also flows in the living system in the form of protons[16,17]. Many microorganisms move by means of a single flagellum. That of bacteria such as *Salmonella* can rotate up to 100 times a second, being driven by a molecular motor attached to its base and embedded in the cell membrane. The motor is entirely powered by the flow of protons (proticity) from one side of the cell membrane to the other. In some marine organisms and organisms living in an alkaline medium, the motor can even be powered by a transmembrane flux of sodium ions[23]. Many other ionic currents flow through the cell and out into the extracellular medium[24].

What I have given above is an oversimplified description of electron transport. In particular, I have been speaking of the outer electrons (again!) as though they were particles localized to the orbit of one molecule or another, which is far from the case.

This is perhaps the place to make a more general point about the scourge of the language we have to communicate in. If we let this language dominate our thoughts, as some philosophers seem to believe we must, it will surely reduce us all to idiots who take everything literally. Actually, we *never* take anything just literally, even in ordinary discourse. If we did, we would not only lose all the meaning behind the words, but also the unfathomable magic. It is just the same in science. As in any attempt to understand, we use whatever tools we have at our disposal to help us think, and good scientific theories are just that - a superior kind of tools for thought. They help us clear our minds in order to receive the greater mysteries of nature. It is in this spirit that I am engaging this whole narrative, the purpose of which is to arrive at an intimate understanding of nature beyond theories and beyond words....

To return to quantum theory, it tells us that the wave-function of each electron (a function of its position weighted by complex numbers, or in standard quantum mechanical language, "a complex-valued function of position") is actually *delocalized* or spread out over the whole system, so that the electron has a finite probability of being found *anywhere* within the system. Furthermore, because the molecules are packed so closely together in a solid state system, the energy levels are no longer discrete, as they are subject, not only to attractive forces from the nuclei, but also repulsive forces from other electrons, as well as vibrations in the molecular lattice. The result is that the outermost electrons are not in discrete energy levels associated with particular atoms or molecules, as they would be in a very dilute ideal gas; but instead, occupy broad bands (of quantum states) delocalized over the lattice, separated in energy by regions where there are no energy levels at all. The lowest energy band involved in bonding - the *valency band* - is filled with electrons. The next higher band is the *conduction band* to which electrons can be promoted by absorbing energy. Electrons in this band are the mobile charge carriers. Between the highest valency level and the lowest conduction level lies the *band gap*, corresponding to the threshold of energy which has to be absorbed to promote a valency electron to a conducting one (see Fig. 7.6). During conduction in a solid state system, similarly, we should not be thinking of charged particles moving one by one around an obstacle course as in a pin-ball machine. Instead, the wave nature of the electron comes into play, so that the charges seem to move much more readily around obstacles than if they were particles, and all kinds of tunnelling effects can occur (literally going 'under' an energy barrier instead of over it), making it necessary to bring in a quantum mechanical description. Readers should bear this in mind whenever and wherever I describe electrons or photons flowing from one place to another. They should automatically add the phrase, 'quantum mechanically', *and* imagine all that the phrase entails. We shall venture further into the strange world of quantum theory in Chapters 10 and 12.

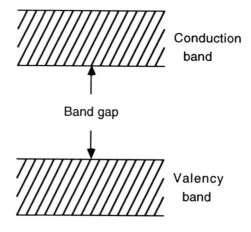

Figure 7.6 The electronic energy bands of the solid state semi-conductor.

The chloroplast and mitochondrial cell membranes are the major sites of charge separation and charge transfer in the living system. There may, however, be other sites associated with macromolecules and macromolecular complexes made up of proteins and nucleic acids. Both theoretical calculations and experimental observations suggest that these too, behave like semiconducting devices when excited by electrical, mechanical or light energy, and can conduct both electricity and proticity[19]. Well-documented quantum tunnelling effects have already been mentioned in connection with the electron and proton transfer proteins in Chapter 3.

The Body Electric Flows and Flows

Thus, electrons (and protons) can flow between molecules and along macromolecules. These local flows are organized or coordinated into spatially more and more extended flow patterns, the entire concerted action of which ultimately enpowers the organism to be alive. Each flow pattern catenates with larger patterns in space and time, and constitutes part of a macroscopic flow which can often be observed in living systems. Indeed, large electric fields are found in tissues and whole organisms[25], which change with injury and anaesthesia. And all developing and regenerating systems have been found to drive steady currents of the order of 1 to 1000mA/cm^2 through themselves[24].

This organized flow of electric currents - meticulously coordinated from the very short range intermolecular charge transfers right up through many intermediate levels of space and time to currents traversing the whole organism - constitutes what I have referred to elsewhere as the coherent electrodynamical field that underlies living organization[26].

Let us briefly revisit Fröhlich's idea of coherent excitation in the light of what has been described so far. I have said that coherent excitations can give rise to long range order in the living system, as well as rapid and efficient energy transfer. We can now see in a more concrete way how that may be achieved. Fröhlich[27] proposes that biological membranes, by virtue of their dipolar structure and the existence of large transmembrane potentials, are particularly prone to such collective vibrational modes. This could explain how the absorption of a single photon, or the binding of a single chemical ligand by a receptor protein could excite hundreds of membrane-bound molecules simultaneously as a first step in the amplification of an external signal arriving on the cell membrane. For all these proteins embedded in the membrane will be vibrating in phase, and hence be in the same state of readiness for receiving the signal. Oscillations in the lipid network could induce simultaneous conformational changes of various proteins anchored in the membrane[16,28].

Proteins are themselves giant dipoles which can undergo coherent dipole excitations over the entire molecule. And in an array of densely-packed giant dipoles such as muscle and the cytoskeleton, the excitation could be coherent throughout the array, accounting for the kind of long-range coordination of molecular machines that is required in biological functioning. For example, McClare[29] suggests that the internal energy of ATP, initially released as a bond-vibration, is resonantly transferred to a pair of oscillators in the actomyosin complex in muscle where the energy is held jointly by the excited pair (excimer) as an 'exciton'. As the exciton is converted into mechanical work, its energy is conducted along the length of the actin filament to activate another ATP molecule, and so on. In that way, the excitation energy is propagated rapidly along the muscle fibre, giving rise to long range coordination of muscle activity. Since then, theoretical and numerical studies suggest that energy from the hydrolysis of ATP could be resonantly transferred via the hydrogen bond to the peptide bond which in turn interacts with the polypeptide lattice nonlinearly to generate a propagating packet of energy, the 'soliton'[30]. Solitons are stable mechanical deformations of the polypeptide chain, in particular, in the α-helical configuration, which can propagate without change in shape or size. Their stability is connected with the competition between two processes: the tendency of the excitation to spread by resonance interaction as against the tendency for the excitation to stay localized due to the mechanical 'softness' or floppiness of the polypeptide chain[31].

Similarly, RNA and especially DNA are also enormous dielectric molecules that can sustain coherent excited modes which may have important biological functions[32], say, in determining which genes are transcribed or translated. We can begin to see how a coherent electrodynamical field makes the organism a vibrant, sensitive whole. In a later chapter, we shall be examining the concept of coherence in the context of quantum theory as it gives important and unusual insights that cannot be obtained otherwise.

In the next two chapters, we shall review some observations and experimental results which lend support to the idea of coherence in living systems, and also present deep challenges to our understanding of the coherent regime in living organisms.

Notes

1. Schrödinger (1944) p.79.

2. Batlogg (1991).

3. Haken (1977).

4. Fröhlich (1968).

5. See Fröhlich (1980). There has been a lot of debate as to whether the 'Fröhlich state' is realizable in practice in living systems. It is thus of great interest that theoretical physicist Duffield (1988) has recently proved that under the most general conditions of energy pumping, the Fröhlich state is globally, asymptotically stable. This means that systems will tend to evolve towards that state, and more over, stay in that state and return to it on being disturbed.

6. See Clegg (1984).

7. See Williams (1980) p. 336-8.

8. McClare (1972) points out that the precision in a chemical system can be greater than 10^{-2}%..

9. See Somogyi et al (1984).

10. c.f. McClare (1972) p.571.

11. See Meggs (1990) and references therein.

12. See also Cope (1975) and references therein.

13. From Clegg (1984) p.R134.

14. McConkey (1982).

15. Redrawn after Clegg (1984).

16. Kell et al (1983).

17. See Williams (1980).

18. Sakurai and Kawamura (1987).

19. See Pethig (1979) Chapter 9. Also Pethig (1992). See also Scott (1984).

20. Szent-Györgi, A. *Introduction to a Submolecular Biology*, Academic Press, New York, 1960, p. 21-22.

21. Thornton, P.R. *The Physics of Electroluminescent Devices*, E. and F.N. Spon Lts., London, 1967.

22. See Tien, H.T. "Membrane photobiophysics and photochemistry." *Prog. Surf. Science* 30 (1/.2)(1989): 1-199.

23. See Meister, M., Caplan, S.R. and Berg, H.C. "Dynamics of a tightly coupled mechanism for flagellar rotation. Baterial motility, chemiosmotic coupling, protonmotive force." *Biophys. J.* (1989)55, 905-14.

24. See Nucitelli (1988) and references therein.

25. Becker (1990).

26. Ho *et al* (1992c).

27. See Frohlich, 1980.

28. Welch *et al* (1982).

29. McClare, 1972.

30. See Scott (1984); and Davydov (1982).

31. See Davydov (1977); also Davydov (1985). I am grateful to Prof. Davydov for sending me his papers.

32. Li *et al* (1983).

CHAPTER EIGHT

HOW COHERENT IS THE ORGANISM?

How to Detect Coherence

There is as yet no *direct* evidence that organisms are coherent, although there are already signs of that from many areas of biological research. Part of the difficulty is that we do not yet know what we should be looking for. In the absence of an appropriate theory, there are no observational criteria which would satisfy the skeptic as to whether a given observation consitutes evidence for coherence. Another difficulty is that until quite recently, there have been very few experiments set up to observe the *living* system, as Schrödinger has already pointed out. Biology has a long tradition of fixing, pinning, clamping, pressing, pulping, homogenizing, extracting and fractionating; all of which has given rise to, and reinforced, a static and atomistic view of the organism. It is no wonder that most biologists still find it difficult to even think of coherence, let alone contemplate how to go about investigating it. As mentioned in the last chapter, it is with the increasing use of sensitive, relatively non-invasive techniques that we shall recognize the animated, sensitive and coherent whole that is the organism.

There are a number of general observational criteria that a coherent system ought to exhibit. I have mentioned them at the beginning of this book, and have dealt with some of them in the earlier chapters: long range order, rapidity and efficiency of energy transduction and transfer, and extreme sensitivity to external cues. Two other criteria are: frequency coupling (in the sense that random energy can become channelled into coherent modes) and fluctuationless functioning of entire populations of molecules. The basis for all these criteria are in the physics of coherence, which will be considered in Chapter 10. For now, I shall review the relevant observations, starting with those in 'conventional' biology and progressing to the more unconventional areas.

Evidence from 'Conventional' Biology

Frequency coupling is well known in biological rhythms, which often show harmonic relationships with one another[1], as for example, the relationship between respiratory rhythm and heart-beat frequency. Similarly, the phenomenon of *sub-harmonic* resonance has been observed in metabolic oscillations, where entrainment of the metabolic rhythm is obtained to driving frequencies which are approximately integer (i.e., whole number) multiples of the fundamental frequency[2]. Recently, a gene has been isolated in *Drosophila*, mutations of which alter the circadian period[3], normally about 24 hours. Remarkably, the wing-beat frequency of its love-song is correspondingly speeded up or slowed down according as to whether the circadian period is shortened or lengthened in the mutant. This correlation spans seven orders of magnitude, linking the circadian period of 10^5s with the period of the love song, which is 10^{-2}s.

A high degree of coordination exists in muscle contraction, as already pointed out in Chapter 1. Insect flight muscle oscillates synchronously with great rapidity, supporting wing beat periods of milliseconds. Many organisms, tissues and cells show spontaneous oscillatory contractile activities that are coherent over large spatial domains with periods ranging from 10^{-1}s to minutes[4]. Similarly, spontaneous oscillations in membrane potentials can occur in a variety of 'non-excitable' cells as well as in cells traditionally regarded as excitable, the neurons, and these oscillations range in periods from 10^{-3}s to minutes, again involving entire cells or tissues (such as the heart, stomach and the intestine). Finally, recent applications of supersensitive SQUID magnetometers to monitor electrical activities of the brain have revealed an astonishing repertoire of rapid coherent changes (in milliseconds) which sweep over large areas of the brain[5]. These findings, and the observations on synchronous firing patterns (40 to 60hz) in widely separated areas of the brain recorded by conventional electrodes, are compelling neurobiologists to consider mechanisms which can account for such long-range coherence[6]. It has been suggested that the synchronization of the oscillatory response in spatially separate regions may serve as 'a mechanism for the extraction and representation of global and coherent features of a pattern', and for 'establishing cell assemblies that are characterized by the phase and frequency of their coherent oscillations'.

It has recently been proven mathematically that synchronization is the rule in any population of oscillators where each oscillator interacts with every other via the absorption of the energy of oscillation, thus resulting in phase locking (that just means they are oscillating together in phase)[7]. As the coupling between oscillators is fully symmetric, i.e., they have completely reciprocal influences on one another, and once they have locked together, they cannot be unlocked (c.f. coupling in chemical systems considered on pp. 44-47). Examples of such phase-locked synchronously oscillating systems may include the flashing of fireflies in various parts of Southeast Asia, the chirping of crickets in unison, as well as the pacemaker cells of the heart, the networks of neurons in the circadian pacemaker and hippocampus, the insulin-secreting cells of the pancreas, and the simultaneously contracting smooth muscle cells in the intestine and the stomach[8]. The coupling energies (or signals) could be visual or auditory in the case of populations of whole organisms, though more subtle electromagnetic interactions cannot be ruled out, especially in the case of cells in tissues (see later).

It has been claimed that muscle contraction occurs in definite quantal steps which are synchronous over entire muscle fibres, and measurements with high speed ultrasensitive instrumentation suggest that the contraction is essentially fluctuationless (as characteristic of a coherent quantum field, see Chapter 10)[9]. Similarly, the beating of cilia in mussels and other organisms also occurs in synchronized quantal steps with little or no fluctuations[10]. This is indeed a collective behaviour which is completely anti-statistical.

Sensitivity of Organisms to Weak Electromagnetic Fields

As consistent with the view, developed in the last chapter, that the flow of electricity at all levels may be responsible for living organization, cells and organisms are sensitive to external electric and magnetic fields. It is known, for example, that nerve cells growing in culture will respond to electric fields as weak as 0.1V/cm - six orders of magnitude below the potential difference that exists across the cell membrane, which is about $10^5 V/cm$. Similarly, skin cells from fish and other animals, and bone cells tend to move towards either the positive or the negative pole in a steady electric field while orienting their long axis at right angles to the field[11]. Orientation is accompanied by shape changes which involve the remodelling of the cytoskeleton. Polymerization of cytoskeletal elements possessing dipole moments will proceed much faster in an electric field[12], which in turn suggests that many, if not all, processes involving changes in cell shape and reorganizations of the cytoskeleton may be associated with endogenous electric fields.

There have been many observations suggesting that diverse organisms are sensitive to electromagnetic fields of extremely low intensities - of magnitudes that are similar to those occurring in nature[13]. These natural electromagnetic sources, such as the earth's magnetic field, provide information for navigation and growth in a wide variety of organisms; while major biological rhythms are closely attuned to the natural electromagnetic rhythms of the earth, which are in turn tied to periodic variations in solar and lunar activities. In many cases (described below), the sensitivity of the organisms to electromagnetic fields is such that they detect signals below the energy level of thermal noise (~kT). This points to the existence of amplifying mechanisms in the organisms receiving the information (and acting on it). Specifically, the living system itself must also be organized by intrinsic electrodynamical fields, capable of receiving, amplifying, and possibly transmitting electromagnetic information in a wide range of frequencies - rather like an extraordinarily efficient and sensitive, and extremely broadband radio receiver and transmitter, much as Fröhlich has suggested (see previous chapter).

Since the 1970s, there have been a lot of additional experimental findings on the biological effects of weak electromagnetic fields[14,15], although perhaps half of the findings are disputed by others. This has become an especially controversial area because of the increasing public concern over the possible harmful effects of all forms of 'man-made' electromagnetic fields[16] associated with high tension power lines, power generators and other installations, as well as various radio-frequency communication devices in the environment. They cover a wide range of frequencies and at intensities at least five orders of magnitude above the natural sources. If our bodies are indeed organized by exquisite electronic music, then these artificial electromagnetic fields may well constitute the worst kind of cacophonous interference.

One major difficulty in reproducing the experimental observations is that the biological effects of given electromagnetic fields may depend crucially on the physiological state or the developmental stage of the systems concerned, and hence, a failure to take that into account will result in a lack of reproducibility. In our laboratory, we have shown that brief exposures of synchronously-developing early *Drosophila* embryos to weak static magnetic fields - during the period when cryptic pattern determination processes are taking place - resulted in a high proportion of characteristic body pattern abnormalities in the larvae hatching 24 hours later (see Fig. 8.1)[17]. The energies of the magnetic fields are well below thermal threshold, so we conclude that there can be no significant effect unless there is a high degree of cooperativity or coherence among the molecules involved in the pattern determination processes reacting to the external fields.

Both static and oscillating magnetic fields have been found to affect biological functions. For example, combinations of static and extremely low frequency oscillating fields, as well as microwave frequency fields modulated by much lower frequencies, cause changes in Ca^{2+} efflux in cells and tissues, and affect the permeability of artificial membrane lipid vesicles to drugs[14,18]. The sensitivity of organisms to electric fields has its basis in the dipolar nature of all molecular and intermolecular interactions, as well as in the ubiquitous charge separation and transfer mechanisms involved in primary energy transductions. The organisms' sensitivity to magnetic fields, however, requires some explanation.

How Can Magnetic Fields Have Biological Effects?

Organisms are in general much more sensitive to weak magnetic fields than to weak electric fields. What is the basis of the organism's sensitivity to magnetic fields, often with infinitesimal amounts of energy? It turns out that magnetic fields have effects which depend less on the energy of the field than on the ability of the field to orientate molecules and atoms with magnetic moments and possibly, on their ability to deflect moving charges.

116

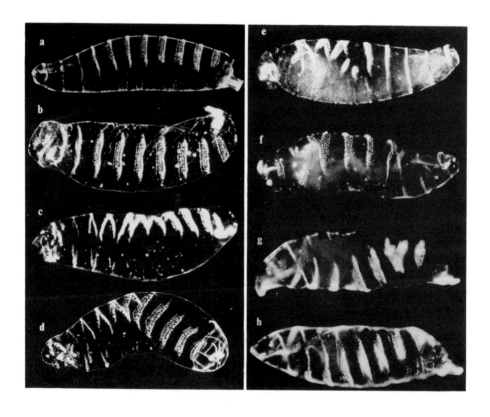

Figure 8.1   Abnormal segmentation patterns in *Drosophila* larvae hatching from embryos exposed to weak static magnetic fields. The consecutive segmental pattern in the normal (a) is converted to various helical or twisted configurations, (b) to (f)[17].

From what has been said in Chapter 6, it is to be expected that time-varying magnetic fields will be able to induce electric currents in the organism, which may be quite effective as organisms are transparent to magnetic fields whereas electric fields are more effectively excluded owing to the build-up of shielding charges in the peripheral layers. Static magnetic fields, however, cannot act by inducing currents but are effective all the same. Static fields act via their ability to orientate magnetically sensitive molecules and by deflecting endogenous electrical currents.

Some molecules can orientate themselves in a magnetic field. As a consequence of their spin, electrons behave as though they are tiny bar magnets each possessing a *magnetic moment,* called the *Bohr magneton,* and is of the order of 10^{-23}J/Tesla, where the Tesla is a unit of magnetic flux density. As we know from Chapter 6, the electrons in an atom are arranged in shells about a central nucleus. A complete shell always contains pairs of electrons with opposite spins. Hence a complete shell has no net magnetic moment. In elements such as iron, cobalt and nickel, there are unpaired electrons, not only in the outer shell but also in an inner shell. In iron, for instance, five of the six electrons in the $n=3$, $l=2$ subshell have parallel spins, so that iron atoms have appreciable magnetic moments and are referred to as *ferromagnetic.* In an unmagnetised sample, the material is characterized by the presence of domains or groups of atoms with all their magnetic moments aligned, but the alignment of different domains are random with respect to one another. On being magnetized, the domains progressively align themselves until all are oriented in the same direction in parallel with the external field. Beyond a certain temperature, the atomic alignment within the domains disappears, and a ferromagnetic material becomes merely paramagnetic (see below). The temperature at which this happens, the *Curie point,* is 770°C for iron.

Atoms and molecules with fewer unpaired electrons do not form domains at ordinary temperatures, and are simply *paramagnetic* or *diamagnetic.* A paramagnetic atom or molecule has a permanent dipole moment, and therefore tends to align itself in the direction of the applied field. A diamagnetic atom or molecule, on the other hand, has no permanent dipole moment; the applied magnetic field affects the orbital motion of the electron in such a way as to produce a magnetic moment in the opposite direction to the applied magnetic field. Pure samples of membrane lipids are known to align themselves in an external magnetic field. It has been suggested that the effect of static magnetic fields on the permeability of lipid vesicles to drugs is due to the summation of diamagnetic alignments of large molecular aggregates to the external field occurring near membrane phase transition temperatures[18].

Could orientation effects be involved in the changes in segmentation pattern of *Drosophila* larvae exposed to static magnetic fields mentioned above? The normal pattern consists of separate consecutive segments, whereas a continuous helical pattern tends to appear in larvae exposed to the magnetic fields (see Fig. 8.1). It must be stressed that this transformed morphology, which I call 'twisted', is highly unusual. I have worked with *Drosophila* for 15 years and never have I, or anybody else come across this kind of body pattern under a variety of pertubations such as heat or cold shocks or exposure to chemicals or solvents[19]. Similarly, tens of thousands of genetic mutants of body pattern have been isolated by *Drosophila* geneticists over the years and so far, not a single one of them has the 'twisted' morphology. This indicates that the static magnetic field exerts its effect in a most specific way, perhaps via orientation of membrane lipids on a global scale.

Membrane lipids (as well as proteins) belong to a large class of molecules called liquid crystals which can exist in a number of *mesophases*, i.e., phases that are neither solid nor liquid, but in between. These mesophases are characterized by long range order in which all the molecules are arranged in a quasi-crystalline arrays. Liquid crystals are easily aligned with electric and magnetic fields - this is the basis of the liquid crystal display screens that come with watches, calculators, computers and computer games[20]. The cell membrane is also known to play a major role in pattern determination[19]. One way in which an external static magnetic field can affect body pattern is if a global alignment of membrane lipids - as a kind of phase transition brought on normally by an endogenous electric field - is involved in pattern determination. In that case, an external magnetic field could easily interact with the endogenous electric field to alter the alignment on a global scale.

Significantly, alternating magnetic fields invariably fail to produce the 'twisted' morphology although other abnormalities are produced in large numbers[21]. This is consistent with the hypothesis that static fields act via the alignment of macroscopic arrays of molecules, which alternating fields are unable to achieve.

A second orientation effect is *Larmor precession*[22] due to oscillating paramagnetic atoms or ions. When exposed to a magnetic field, the oscillating magnetic atom tries to align itself, and in so doing, precesses like a top around the direction of the magnetic field; the frequency of precession (the Larmor frequency) being proportional to the intensity of the magnetic field. This is an instance where the effect could appear at low field intensities and disappear at high intensities. Thus, the effect of electromagnetic fields on the movement of the paramagnetic ion, Ca^{2+}, across the cell membrane exhibits both frequency and intensity windows[14].

Another orientation phenomenon is well documented in chemical reactions[23]. This occurs in reactions where a covalent bond is split, resulting in 'free radicals' each carrying a single unpaired electron. As the covalent bond was formed originally by two electrons of opposite, or antiparallel, spins, the radicals resulting from the reaction will also have antiparallel spins, and is said to be in the *singlet* state, as both states have the same energy level,

$$R_1\text{-}{\downarrow}{\uparrow}\text{-}R_2 \rightarrow R_1{\downarrow} + R_2{\uparrow}$$

However, it often happens that the separated radicals may have parallel spins (that is, with the arrows in the separated radicals pointing in the same direction), referred to as the triplet state, as they will populate three different energy levels in a magnetic field (see below). This would cause them to drift apart so that one or the other of the radicals can react with another molecule yielding another pair of free radicals also with parallel spins. In order for covalent bonds to form, the spins must be antiparallel, and in the absence of an external magnetic field, all the triplet states have approximately the same energies so the relative spins soon change and bond formation will take place.

In the presence of an external field, however, the spins align themselves so that they are both either in the direction of the field, T_{+1}, or in the opposite direction, T_{-1}, or they can be antiparallel in the field direction, T_0. These different orientation states no longer have the same energies, and as the magnetic field increases in intensity, they diverge more and more (see Fig. 8.2). The $T_{\pm 1}$ states become trapped, and can no longer get back to the antiparallel configuration which is necessary for forming covalent bonds. Hence the reaction rate in the presence of the magnetic field will be greatly diminished. By contrast, oscillating or alternating magnetic fields will be expected to enhance reactions rates, as they facilitate the changes in relative spins. Here is another instance where static magnetic fields will have very different effects from oscillating magnetic fields.

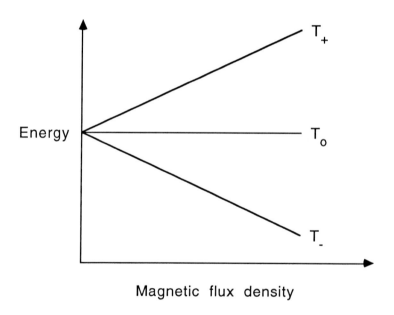

Figure 8.2 The separation of the triplet energy levels due to different alignments of the magnetic free radical pairs as the magnetic flux density increases.

Finally, static magnetic fields will deflect endogenous electric currents flowing at right angles to the direction of the imposed field. During early embryogenesis in *Drosophila,* when pattern determination is taking place, considerable electrical activities are also evident (see next chapter), quite apart from the transembryonic ionic currents which have been found flowing through and around all developing organisms that people have looked at. The ionic currents have a drift velocity far too slow to be appreciably deflected by the external magnetic field[17]. Furthermore, it is difficult to see how the deflection of ions in free solution can affect the cell membrane and influence pattern determination. If there is an electric current which is sensitive to the magnetic field in the manner suggested by the specific 'twisted' transformations, it would have to flow along the surface of the cell membrane and be coherent throughout the entire surface. There is no evidence for the existence of such a current, although as mentioned in Chapter 7, it is possible for electric currents - in the form of electrons or protons - to flow along membranes.

We have only scratched the surface of coherence in living organization. The sensitivity of organisms to weak magnetic fields, strange as it may seem, can in principle be explained in terms of classical physicochemical phenomena which are relatively well understood. In the next chapter, I shall describe some novel observations in living organisms that have as yet no definite analogy in non-living systems.

Notes

1. Breithaupt (1989).
2. Hess (1979).
3. Kyriacou (1990).
4. Berridge *et al* (1979).
5. Ribary *et al* (1991).
6. Gray *et al* (1989).
7. See review by Stewart (1991) and references therein.
8. See also Winfree (1980).
9. Iwazumi (1979).
10. Baba (1979).
11. See Ferrier (1986) and references therein.
12. See Meggs (1990) and references therein.

13. Presman (1970).

14. Adey (1989).

15. Becker (1990).

16. For details and experimental observations , see Ho *et al* (1993).

17. See Ho *et al* (1992a).

18. Liburdy and Tenforde (1986).

19. See Ho (1987) and references therein.

20. See special issue on liquid crystals, *Physics Today*, May, 1982.

21. Ho, M.W. and French, A., unpublished results.

22. See Edmonds (1992).

23. See McLauchlan (1992) .

CHAPTER NINE

LIFE IS ALL THE COLOURS OF THE RAINBOW IN A WORM

Light and Living Matter

Light and matter are intimately interrelated. The first inkling of that comes from observations showing how, at the quantum level, both light and matter exist as wave and particle (see Chapter 6). I shall deal with the wave-particle duality of light and matter more substantially in the next chapter. In this chapter, I describe a number of experimental observations suggesting that light and *living* matter have such a special relationship that it pushes at the very frontiers of current research in quantum optics and other nonlinear optical phenomena in condensed matter physics.

The Light That Through The Green Fuse

German quantum physicist turned biophysicist, Fritz Popp, and his group in Kaiserslautern, have been investigating light emission from living organisms for nearly 20 years. They find that practically all organisms emit light at a steady rate from a few photons per cell per day to several hundred photons per organism per second[1-3]. The emission of *biophotons*, as they are called, is somewhat different from well-known cases of bioluminescence which occurs in isolated species such as fireflies and luminescent bacteria, for example. Biophoton emission is universal to living organisms, occurring at intensities generally many orders of magnitude below that of bioluminescence, and in contrast to the latter, is not associated with specific organelles. Nevertheless, biophoton emission is strongly correlated with the cell cycle and other functional states of cells and organisms, and responds to many external stimuli or stresses. The response to temperature is highly nonlinear: a sharp increase in emission rate as temperature rises followed by oscillations before decaying back to a steady level. Analyses of the emitted light reveal that it typically covers a wide band (wavelengths from about 250 to 900nm) around the optical range, with an approximately equal distribution of photons throughout the range, i.e., the energy levels are approximately equally populated. This distribution deviates markedly from that characteristic of systems at thermodynamic equilibrium, giving us yet another indication that the living system is far, far away from thermodynamic equilibrium in its energy profile, as already pointed out in Chapters 4 and 5.

Biophotons can also be studied as stimulated emission after a brief exposure to light of different spectral compositions (different combinations of frequencies). It has been found, without exception, that the stimulated emission decays, not according to an exponential function characteristic of non-coherent light, but rather to a hyperbolic function (see Fig. 9.1) which Fritz Popp claims to be a sufficient, or diagnostic, condition for a coherent light-field[1,2]. What this implies is that photons are held in a coherent form in the organism, and when stimulated, they are emitted coherently, like a very weak, multimode laser. (Such a multimode laser has not yet been made artificially, but it is at least not contrary to the theory of coherence in quantum optics as developed especially by Glauber[4].) By 'photons', I include also packets of electromagnetic energies below the visible range; it could in principle extend all the way to the radio frequencies and extremely low frequency (ELF) end of the spectrum.

The hyperbolic function takes the general form,

$$x = A(t + t_0)^{-1/\delta}$$

where x is the light intensity, A, t_0 and δ are constants, and t is time after light exposure. The equation expresses the characteristic that light intensity falls off as a power of time, pointing to the existence of memory in the system.

The phenomenon of hyperbolic decay can be intuitively understood as follows. In a system consisting of non-interacting molecules emitting at random, the energy of the emitted photons are completely lost to the outside, or converted into heat, which is the ultimate non-coherent energy. If the molecules are emitting coherently, however, the energy of the emitted photons are not completely lost. Instead, part of it is coupled back coherently, or reabsorbed by the system (rather like the interaction between coupled oscillators described in the previous chapter). The consequence is that the decay is delayed, and follows typically a hyperbolic curve with a long tail. Other non-linear forms of delayed decay kinetics, such as oscillations, are predicted for a coherent field, and are also often observed.

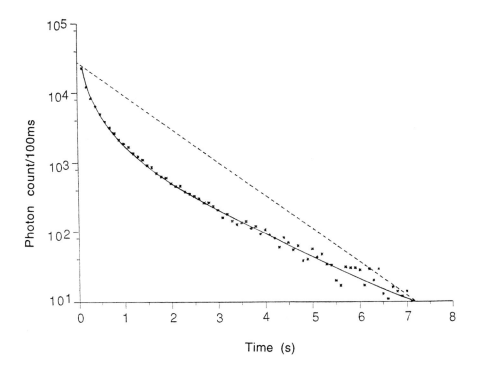

Figure 9.1 Hyperbolic decay kinetics of stimulated light emission from a batch of synchronously developing early Drosophila embryos[5].

The hyperbolic and nonlinear decay kinetics is uniform throughout the visible spectrum as shown by analyzing the spectral composition of the emission stimulated by monochromatic light or light of restricted spectral compositions[3,6]. The stimulated emission always covers a broad spectrum regardless of the composition of the light used to induce it, and furthermore, can retain its spectral distribution even when the system is perturbed to such an extent that the emission intensity changes over several orders of magnitude. These observations are consistent with the idea that the living system is one coherent 'photon field' which is bound to living matter. This photon field is maintained far from thermodynamic equilibrium, and is coherent simultaneously in a whole range of frequencies that are nonetheless coupled together to give, in effect, a single degree of freedom. This means that random energy input to any frequency will become delocalized over all the frequencies.

It must be stressed that 'a single degree of freedom' is maintained only as a statistical average. As will be made clear in the description of quantum coherence in the next chapter, coherence does not mean uniformity, or that every part of the organism must be doing the same thing or vibrating with the same frequencies. There can indeed be domains of local autonomy such as those that we know to exist in the organism. Furthermore, as organisms have a space-time structure, any measurement of the degree of freedom performed within a finite time interval will deviate from the ideal of one, which is that of a fully coupled system with little or no space-time structure. Another source of variation may arise because some parts of the system are temporarily decoupled from the whole, and the degree of coherence will reflect changes in the functional states of the system. Such a variation in the degree of coherence appears to be associated with malignant tumour cells.

Long-Range Communication between Cells and Organisms

In considering the possibility that cells and organisms may communicate at long range by means of electromagnetic signals, the Soviet biologist, Presman[7] points to some of the perennial mysteries of the living world: how do birds in a flock, or fish in a shoal, move so effortlessly in unison? During an emergency, the speed and strength of action of the organism are much greater than the normal working level. The motor nerve to the muscle conducts at 100 times the speed of the vegetative nerves that are responsible for activating processes leading to the enhancement of the contractile activity of the muscles required in a crisis: adrenalin release, dilatation of muscular vessels and increase in the heart rate. Thus, it appears that the muscle receives the signals for enhanced coordinated action long before the signals arrive at the organs responsible for the enhancement of muscle activity! This suggests that there may be a system of communication that sends emergency messages simultaneously to all organs, including those perhaps not directly connected with the nerve network. The speed with which this system operates seems to rule out all conventional mechanisms. (This is also true for visual perception and muscle contraction described in Chapter 1.) Presman proposes that electrodynamical signals are involved, which is consistent with the sensitivity of animals to electromagnetic fields. Electrodynamical signals of various frequencies have been recorded in the vicinity of isolated organs and cells, as well as close to entire organisms[8]. In our laboratory, we have recorded profuse electrical signals from fruitfly embryos (1hz to 30hz, 1hz = 1cycle per second) during the earliest stages of development when the embryos are not yet cellular[9]. Whether these signals are involved in communication is not yet known.

The photon-emission characteristics of normal and malignant cells have been investigated[10]. While normal cells emit less light with increasing cell density, malignant cells show an exponential increase in light emission with increasing cell density. This suggests that long-range interactions between the cells may be responsible for their differing social behaviour: the tendency to disaggregate in the malignant tumour cells as opposed to attractive long range interactions between normal cells. The difference between cancer cells and normal cells may lie in their communicative capability which in turn depends on their degree of coherence. The parameter $1/\delta$ in the hyperbolic decay function (see above) can be taken as a measure of incoherence, as it is directly correlated with the inability of the system to re-absorb emitted energy coherently. This parameter is shown to increase with increasing cell density in the malignant cells, and to *decrease* with increasing cell density in normal cells.

Similar long range interactions between organisms have been demonstrated in *Daphnia* where the light emission rate varies periodically with cell number in such a way as to suggest a relationship to the average separation distances between individual organisms which are submultiples of the body size - of about 3 mm (see Fig. 9.2)[11].

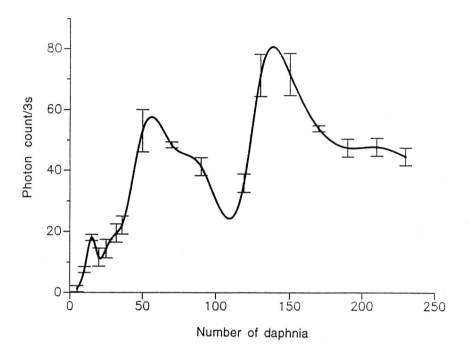

Figure 9.2 Spontaneous light emission in *Daphnia* as a function density[12].

Finally, in synchronously-developing populations of early *Drosophila* embryos, we have recently discovered the remarkable phenomenon of super-delayed luminescence in which intense, often prolonged and multiple flashes of light are re-emitted with delay times of 1 minute to 8 hours after a single brief light exposure. Some examples are presented in Figure 9.3[13]. The phenomenon depends, among other factors, on the existence of synchrony in the population. Although the timing of light exposure must fall within the first 40 minutes of development in order to obtain superdelayed luminescence, the occurrence of the flashes themselves do not obviously

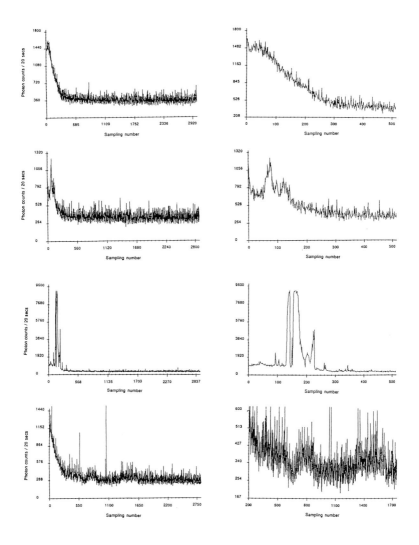

Figure 9.3 Superdelayed luminescence in *Drosophila*[13]. Continuous recordings of aggregated photon count per 20s from synchronously-developing embryos. Top trace, controls, not exposed to light. The other traces are of embryos previously exposed to white light for 1 min. and exhibitng different forms of superdelayed luminescence. Traces on the right are expanded versions of those on the left.

correlate with specific embryonic events. The flashes, therefore, give information concerning the physical state of the embryos at the time of light stimulation - such as the existence of a high degree of coherence - rather than at the time during which the flashes themselves occur. Superdelayed re-emission takes many forms. It may consist of a single sharp burst of light (< 20s) or of several sharp bursts separated by long periods of background emission. It may also come in a single prolonged burst or a series of prolonged bursts. The duration of each prolonged burst varies from 1 to 30 minutes. For the multiple prolonged flashes, the total luminescence produced - in excess of the background emission - shows significant first and second order dependence on the number of embryos present. All these observations suggest that superdelayed luminescence results from cooperative interactions among embryos within the entire population, so that most if not all the embryos are emitting light simultaneously.

This implies that each embryo has a certain probability of re-emission after light stimulation, so that it can either trigger re-emission in other individuals or alternatively, its re-emission could be suppressed by them. Only when the population is re-emitting at the same time is the intensity sufficient to be registered by the photon-counting device. On the other hand, re-emission in the entire population could also be suppressed, such that in approximately 30 to 40% of the cases, there is no clear indication of superdelayed re-emission even when all the necessary criteria have been satisfied.

The phenomenon bears some resemblance to various nonlinear quantum electrodynamic effects (see next Sec.). As yet, we do now know whether any functional significance could be attached to it. *Drosophila* females typically lay eggs just before sunrise, so the external light source could be used as an initial synchronizing signal or *Zeitgeber*, which maintains the circadian and other biological rhythms. The superdelayed re-emission could then be a means of maintaining communication and synchrony among individuals in the population. On the other hand, the flashes may simply be the embryos' way to inform us of their globally coherent state at the time when light stimulation is applied. This enables the embryos to interact nonlinearly to generate light emission that is coherent over the entire population, and orders of magnitude higher than the spontaneous emission rate.

Where Do Biophotons Come From?

Let us consider some possibilities regarding the source of biophotons. I have suggested that single embryos have the probability of re-emission, and indeed, we have succeeded in obtaining an image of single *Drosophila* embryos built up by integrating the photons from their superdelayed luminescence[13]. The results indicate that the light is emitted over the entire embryo, though certain areas may be emitting more strongly than others at different times. The dynamics of emission can be quite complex, as we were able to demonstrate in video recordings of the image in real time, which we subsequently succeeded in obtaining - with much thanks to help from Patricia Thomkins and Paul Gibson of Photonics, U.K.[14]. This lack of specific localization of the light emitted is consistent with its broad spectrum, and also with *a priori* expectations as explained below. We are in the process of analyzing the spatial-temporal characteristics of the stimulated light emission, which promises to be quite fascinating.

Light is generally emitted from an excited atom or molecule, when an electron in the outermost shell, having been promoted to an excited energy level by, say, a collision with another molecule or absorption of energy by other means, falls back into a lower energy level. Light emission does not always occur, however. The excited electron can often start to move, thus becoming an electric current, or it can be involved in a chemical reaction as explained in the previous chapter. The electron can also relax back to the ground state non-radiatively, that is, instead of emitting photons, it can give off energy as phonons (sound waves), or as heat. In the case of light emission, the energy of the emitted photon will be equal to the difference between the energy levels of the excited and the ground state, which determines the frequency of the photon emitted,

$$E_e = E_1 - E_0 = h\nu$$

where E_e, E_1 and E_0 are respectively, the energies of the photon emitted, the excited level, and the ground level, h is Planck's constant, and ν is the frequency of the emitted photon.

This will prove to be an oversimplified account, as photons are even more intimately involved with matter than the impression I have given so far. In quantum electrodynamics, it is supposed that the orbital electrons and the nucleus are exchanging 'virtual' (i.e., not quite real) photons all the time, which is why the electrons can manage to 'move around in their orbits' without radiating electromagnetic energy outside[15].

In a solid state system, as explained in Chapter 7, outer electrons are neither localized to single atoms or molecules, nor do they have single energy levels. Instead, continuous and delocalized bands of frequencies are separated by band gaps. The low energy valency band, filled with electrons, is separated by a band gap from the conduction band containing mobile electrons to which electrons can be promoted by absorbing energy such as heat or light. If the excited electron is not conducted away, it may relax back to the ground state (to recombine with the positively charged 'hole' left behind) by radiating a photon, or non-radiatively as described above. Another source of photons in a solid state system is from 'excitons', i.e., an excited electron-hole pair which can propagate over long distances within the system before giving up the energy by emitting a photon.

As we have seen, charge separation underlies the primary bioenergetic transduction processes associated with biological membranes, and the formation of excitons and their propagation is believed to be widely involved in energy transduction and in biocommunication. Rattemeyer and Popp[16] suggest that the DNA molecule is an excited duplex, or *exciplex*, in which photons are effectively stored between the two DNA strands, and hence can be a source of emitted biophotons. Exciplex formation in DNA has been shown to predominate even at room temperature[17]. So it is not surprising that living systems could emit light from processes taking place all over the cell. And on account of the solid state nature of the cell, one must also suppose that since all the electrons are delocalized, so too, must all the photons in the system be delocalized. Therefore, it is probably misguided and quite fruitless to try to identify from which specific chemical reactions biophotons originate. However, it is one thing to suggest that biophotons come from all over the cell or organism which constitutes a coherent photon field, and quite another to imagine how coherence would manifest itself in light emission. Or, put the other way round, it remains a problem to work out precisely what the characteristics of light emission are telling us about coherence in the living system.

Spontaneous emission is generally too weak to be analyzed in detail. There is little possibility of observing interference effects directly - which are characteristic of coherent monochromatic light (see next chapter) - for biophotons clearly are not monochromatic even though they can be coherent, as the theory of quantum optics[4] explicitly allows for coherence in a wide band of frequencies which are coupled together. More opportunity presents itself in the analysis of stimulated light emission. As mentioned earlier, the hyperbolic decay kinetics is considered by Popp as a sufficient condition for a coherent field. This is consistent with the observation that the hyperbolic decay kinetics is uniform over the entire optical spectrum, and that random energy of any frequency fed into the system is rapidly delocalized over all frequencies, implying that all the frequencies are coupled in a single mode as required for coherence[4].

In the case of superdelayed luminescence in *Drosophila* embryos, we have to explain both the extraordinarily long delay in re-emission, and the greatly enhanced re-emission rate. There are two kinds of novel electrodynamical phenomena in physics which bear certain formal similarities to superdelayed luminescence in our biological system. (The similarity is formal only because the long time constants - tens of minutes and even hours before re-emission - involved in the biological systems have as yet no equals in physical systems.) One of these, cavity quantum electrodynamics, is the control of spontaneous radiation from excited atoms in resonant cavities[18,19]; the other is the localization of light in semi-conducting material containing dielectric microstructures[20]. Both physical phenomena emphasize the quantum delocalization of electromagnetic energy and collective interactions on which biological organization depends.

Cavity quantum electrodynamics is associated with experiments demonstrating that spontaneous radiation from excited atoms can be greatly suppressed or enhanced by placing them in a special reflecting cavity which restricts the modes that the atom can radiate into. Strong coupling between the atoms and the radiation field within the cavity can then lead either to a suppression of spontaneous emission of excited atoms or its great enhancement, corresponding to the *subradiant* and *superradiant* modes, respectively. The early embryo can, perhaps, be regarded as a cavity resonator in which the spontaneous emission of atoms coherently excited by light exposure has become suppressed for various periods of time before they become re-emitted in a greatly enhanced rate, possibly as the result of some natural change of state as development proceeds subsequent to light stimulation. Or, as mentioned above, the re-emission can be suppressed indefinitely, in the subrandiant mode. (The light exposure has no deleterious effects on development, and qualifies as a non-invasive probe.) The situation is further complicated by our not dealing with one single cavity resonator, but rather with a population of nearly identical resonators which can further interact to emit cooperatively and simultaneously, over the entire population.

The general formal resembances between superdelayed luminescence and cavity quantum electrodynamics include the characteristic multiple emission peaks - the most frequently observed emission pattern, the stochastic nature of the delay times, and the size and shape of the peaks. The dramatic multiple peaks occur between half to one-third of the time. The next most frequent forms are single long-lasting intense peaks, and single to multiple sharp bursts of light (each less than 20s duration). In some cases, no re-emission at all is observed, even though all the controllable parameters such as age and degree of synchrony of the embryos are arranged to be the same. This is consistent with the observation in quantum cavity electrodynamics that the geometry of the cavity (or in our case, the precise distribution of the embryos which are placed in the quartz cuvette by the mother flies themselves) can affect whether superradiance, or its converse, subradiance occurs. When superradiance occurs, the light also tends to be directional, which means that not all of the emissions may be detected by the photomultiplier placed in one definite orientation. In order to investigate superdelayed luminescence and the more usual delayed re-emission properly, we are currently developing direct low light level imaging of single embryos as well as populations of embryos.

Localization of light is a related quantum electrodynamic effect which occurs as the result of coherent scattering and interference in semiconducting materials containing dielectric microstructures which are close to the dimensions of the wavelength of light that is being scattered, i.e., $a \sim \lambda/2\pi$. Under those conditions, there are no propagating modes in any direction for a band of frequencies. Any 'impurity' atom - i.e., atoms different from the bulk dielectric medium - with a transition frequency in this band gap will not exhibit spontaneous emission of light. Instead, the emitted photon will form a bound state to the atom. In other words, the photon will be trapped indefinitely in the material. When a collection of 'impurity' atoms is present in the dielectric, a single excited atom can transfer its bound photon to neighbouring atoms by resonant dipole-dipole interactions. The distance for such photon tunneling is approximately ten times the dimension of the microstructures. Thus, a photon-hopping conduction results, involving a circulation of photons among a collective, which greatly increases the likelihood of coherent stimulated emission or laser action.

The conditions for light localization and subsequent coherent stimulated emission may well be present in the early embryo. There are many candidates for dielectric microstructures of the dimensions of the wavelengths of light, and, on account of the molecular diversity of the biological system, there is no shortage of species acting as so-called 'impurities'. For example, globular proteins range from 5 to 10 nm; protein complexes and ribosomes, 20 to 30nm. These can all contribute to the localization of various frequency bands of photons. As the embryo develops, however, conditions favourable for stimulated coherent emission of the trapped photons may become realized. In connection with the trapping or storage of photons, it has been observed that death in organisms invariably begins with a sharp increase in the intensity of light emission (at least 3 orders of magnitude above the self-emission rate) which can persist undiminished for more than 48 hours[21,22].

The above phenomena of light emission from living organisms are so unusual that they would certainly be dismissed as curiosities or artifacts were it not for the existence of a general theory of coherence in biological systems within which these phenomena could begin to be understood. There is another deeper reason why new theories in general are important for science:

they direct us to new observations which may very well not be made otherwise. This has been my experience in years of scientific research in many areas. All of the most recent discoveries in my laboratory have been inspired by the theory of coherence; in particular, I was very interested in the role played by coherence in body pattern determination[7,13], which motivated my collaborative work with Frtiz Popp in the first place. Our subsequent findings on light emission in *Drosophila* embryos encouraged me to explore further the optical properties of the developing embryo. This has led to more exciting discoveries, I shall describe one of these.

Life is All the Colours of the Rainbow in a Worm

For quite some time, it had occurred to me that coherence in the organism ought to be reflected in its molecular organization, so one should be able to see evidence for that using various optical techniques, such as the polarizing microscope, which people have routinely used for studying crystalline material in rocks and various 'inert' biological material such as fibres, bones, teeth and so on, and more recently, to investigate phase ordering in liquid crystals. In a few instances, it has even been used on isolated, contracting muscle fibres. If organisms are indeed special solid state systems with the quasi-crystalline structure of liquid crystals, then there is every reason that the polarizing microscope may reveal the regimes of dynamic order in the molecular arrangement of its cells and tissues. As I was most interested in the early stages of pattern determination, I had expected to see some kind of 'prepatterning' in the early embryo which would mirror the body pattern that develops overtly much later on. This idea remained at the back of my mind, but I never seriously pursued it because we did not have a polarizing microscope in the Department, and there were many other things to do besides.

As luck would have it, Michael Lawrence, a colleague who worked as a microscopist and designer in the BBC, came over to visit one day some months ago, and told me he was filming crystallization under the polarizing microscope in the Department of Earth Sciences. I pursuaded him that we ought to look at my *Drosophila* embryos. He agreed to do so almost immediately. And as a result, we discovered a new imaging technique that enables us to see all the colours of the rainbow in a living, crawling, first instar *Drosophila* larva.

To see it for the first time was a stunning, breathtaking experience, even though I have yet to lose my fascination for it after having seen it many, many times subsequently. The larva, all of one millimeter in length and perfectly formed in every minute detail, comes into focus on the colour-TV monitor as though straight out of a dream. As it crawls along, it weaves is head from side to side flashing jaw muscles in blue and orange stripes on a magenta background. The segmental muscle bands switch from brilliant turquoise to bright vermillion, tracking waves of contraction along its body. The contracting body-wall turns from magenta to purple, through to iridescent shades of green, orange and yellow. The egg yolk, trapped in the ailmentary canal, shimmers a dull chartreuse as it gurgles back and forth in the commotion. A pair of pale orange tracheal tracts run from just behind the head down the sides terminating in yellow spiracles at the posterior extremity. Within the posterior abdomen, fluorescent yellow malpighian tubules come in and out of focus like decorative ostrich feathers. And when highlighted, white nerve fibres can be seen radiating from the ventral nerve cords. Rotating the microscope stage 90° causes nearly all the colours of the worm instantly to take on their complementary hues. It is hard to remember that these colours have physical meaning concerning the shape and arrangements of all the molecules making up the different tissues.

It was some time before we realized that we have made a new discovery[23]. The technique depends on using the polarizing microscope unconventionally, so as to optimize the detection of diverse, small birefringences or anisotropies in the molecular structures of the tissues[24,14]. We do not yet know how to interpret the colours precisely in terms of molecular structures, except that as with ordinary crystalline and liquid crystalline material, they do tell us about long-range order in arrays of molecules. The full colours only appear under our conditions, and leads us to think we are picking up phase ordering in biological molecules in living organisms that have never been observed before, and cannot be observed under conventional conditions. Furthermore, we have accomplished the first ever, high resolution and high contrast imaging of an entire, living, moving organism. Finally, the very idea of using polarizing light microscopy to look at *dynamic* order *within* the organism is also new.

What is so suggestive of dynamic order is that the colours wax and wane at different stages of development. The early stages associated with pattern determination are some of the most colourful ones. (Although I could see nothing of the prepatterning that I had expected, I am far from disappointed. It confirms my belief that the best experiments always tell one something one has not quite thought of before, as they are acts of communicating with nature. More about that in Chapter 11.) The most dramatic change is in the final stages when the colours intensify in the embryos that have started to move several hours before they are due to hatch. It is as though energy, say in the form of an electric field, is required to order the molecules, so that even though the structures are formed, the molecules are not yet coherently arranged, and requires something like a phase transition to do the job. The closest analogy one could think of is the phase ordering of liquid crystals in electric and magnetic fields, as biological membranes and muscle fibres in particular, have properties not unlike those of liquid crystals (see previous chapter). As consistent with the above interpretation, the colours fade when the organism dies or becomes inactive due to dehydration or to the cold. In the latter cases, the colours return dramatically within 15 minutes when the organism is revived.

It is also highly significant that in all live organisms examined so far, which includes the *Drosophila* larva, *Daphnia*, rotifers, nematodes, larva of the brine shrimp, hatchling of the zebra fish and the crested newt, the anterior-posterior body axis invariably corresponds to the major polarizing axis of all the tissues in the entire organism. This is a further indication that some global orienting field is indeed responsible for determining the major body axis.

This technique works on all live biological tissues. Tissues which have been fixed and stained fail to show any colours. Thus, we seem to have a technique for imaging dynamic order which is correlated with the energetic status of the organism. Is this evidence of the non-equilibrium phase transition to dynamic order that Fröhlich and others have predicted on theoretical grounds?

All our investigations on the optical properties of living organisms are done using relatively non-invasive techniques that, as mentioned in earlier chapters, enable us to study living organization in the *living, organized* state. That is in itself a major motivation towards the development of non-invasive technologies. Ultimately, they lead to a different attitude, not only to scientific research or knowledge acquisition (see Chapter 11), but to all living beings. For as these technologies reveal both the immense subtlety and exquisite sensitivity of biological organization, they engender an increasingly sensitive and humane regard for all nature.

There is neither the time nor the place to go into details about all the new observations and discoveries that have been made and continue to be made in the area of electrodynamical coherence, both in my own laboratory and in other laboratories. Nevertheless, I hope to have conveyed some of the excitement and fruitfulness in thinking about the physics of organisms.

Notes

1. See Popp (1986). The first studies on biophoton emission were actually carried out by the Soviet scientist, Alexander Gurwitsch and his sister and co-worker, Anna Gurwitsch, who has continued the investigations up to the present-day. They showed that certain frequencies of uv radiation from dividing cells were 'mitogenic' in that they could stimulate otherwise resting cells to divide. This research is currently undergoing a revival in the hands of Russian developmental biologist, Lev Beloussov. See Gurwitsch (1925); also A. Gurwitsch (1992) in Popp *et al* (1992). I am very grateful to Lev Beloussov for unpublished information on his efforts to extend Gurwitsch's observations to developing embryos.

2. Popp and Li (1992).

3. Popp, *et al* (1981).

4. Glauber (1969).

5. From Ho *et al* (1992c).

6. Musumeci *et al* (1992)

7. See Presman (1970).

8. Pohl (1983).

9. See Ho *et al* (1992b).

10. Schamhart and van Wijk (1986).

11. Galle *et al* (1991).

12. From Ho and Popp (1993).

13. See Ho (1992c).

14. Video sequences shown in public in the 1993 International Science Festival, April 14, Edinburgh.

15. See Richard Feynman's fascinating little book, *Q.E.D. The Strange Theory of Light and Matter*.

16. Rattemeyer and Popp (1981).

17. Vigny and Duquesne (1976).

18. Gross and Haroche (1982).

19. Haroche, and Kleppner (1989).

20. John (1991).

21. Li *et al* (1983).

22. Neurohr, R. unpublished observation (1989).

23. See Ho and Lawrence (1993a; 1993b).

24. See Ho, Lawrence and Saunders (1993).

CHAPTER TEN

QUANTUM ENTANGLEMENT AND COHERENCE

What *is* Coherence?

A key notion in the new perspective of living organization developed in the previous chapters is 'coherence'. Coherence in ordinary language means correlation, a sticking together, or connectedness; also, a consistency in the system. So we refer to people's speech or thought as coherent, if the parts fit together well, and incoherent if they are uttering meaningless nonsense, or presenting ideas that don't make sense as a whole. Thus, coherence always refers to wholeness. However, in order to appreciate its full meaning and more importantly, the implications for the living system, it is necessary to make incursions into its quantum physical description, which gives insights that are otherwise not accessible to us.

Before we do that, we must venture more deeply into the mysterious world of quantum physics. We have already been using a good deal of quantum physics in previous chapters, as when dealing with statistical mechanics, the structure of atoms, the physics of solid states, and so on, which shows how important it is for understanding contemporary physics. Here we shall delve into the fundamental nature of physical reality revealed to us by quantum theory, and we shall do so with the help of two kinds of observations, which illustrate most clearly how the quantum world is completely at odds with the classical world of Newtonian mechanics. The first kind of observations demonstrates the *wave-particle duality* of physical reality, and the second, the phenomena of *nonlocality* and *entanglement*. These experiments also impinge upon the issue concerning the role of the observer in science. Within the classical framework, the role of the observer is strictly external to the system observed and it is supposed that no influence passes from the one to the other. By contrast, the observer and observed in the quantum world seem somehow inextricably entangled. As we shall see, quantum entanglement has a lot to do with quantum coherence.

Wave-particle Duality

We have seen in Chapter 6 that even within classical physics, light exhibits properties which are consistent with its being composed of waves or of particles. However, in quantum physics, we must think of light as *simultaneously* wave-like and particle-like. Most quantum physicists have given up trying to say what it really is by calling this property (which belongs to both light and matter), *the wave-particle duality*. It manifests itself most graphically in one of the earliest quantum-mechanical experiments.

In the simplest version of this experiment, a beam of monochromatic light is fired through a screen containing a pair of narrow slits onto a photographic plate (see Fig. 10.1). When only one of the slits is opened, an image of

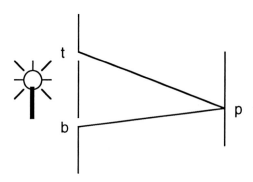

Figure 10.1 The two-slit experiment. (See text)

the slit is recorded on the photographic plate, which, when viewed under the microscope, would reveal tiny discrete spots (see Fig. 10.2). And this is consistent with the interpretation that individual particle-like photons, on passing through the slit, have landed on the photographic plate, where each photon causes the deposition of a single silver grain.

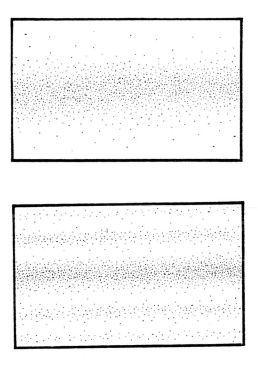

Figure 10.2 The exposure pattern on the photographic plate. Top, with one slit open; bottom, with both slits open.

When both slits are opened, however, an interference pattern arises on the photographic plate, which is consistent with a wave-like behaviour of the light: the two wave trains on passing through the slits, arrive at different parts of the screen either in phase, when they reinforce each other to give a bright zone, or out of phase, so that they cancel out each other to give a dark zone. On examining the photographic plate under the microscope, however, the same graininess appears, as though the light waves become individual particles as soon as they strike the photographic plate (Fig. 10.2).

The result is the same even if the light intensity is reduced to such an extent that only individual photons will pass through the slits and arrive on the screen *one at a time*. This means that each photon has somehow passed through both slits and interfered with itself, i.e., each particle behaves in a wave-like way entirely on its own in being able to pass through both slits. Yet on encountering the photographic plate, it becomes a particle again! If one tries to be clever and place a photon-detector at one of the slits so that the observer can tell which slit the photon has passed through, then the interference pattern will disappear. In order for interference to take place on the photographic plate, it seems that we must remain 'ignorant' as to which slit the photon has 'actually' passed through. The photons retain both alternatives of going through the top and the bottom slits. Similar experiments have been done with electrons, or recently, with neutrons, which are 1800 times as massive as the electron, and the results tell us the same thing. The neutron can be 'split' into two (by a suitable beam splitter), and on recombining gives an interference pattern exactly as the single photon does. Again if we were to put detectors in either path, then we would always find the whole neutron in either one or the other, but never in both simultaneously.

The way quantum mechanics explains this result is to 'say' mathematically that the photon has a 'probability amplitude' - expressed as a complex number, of passing through the top or the bottom slit. These amplitudes express the quantum mechanical alternatives available to the photon. But they are not probabilities in the classical sense. In order to get the correspondence to classical probabilities (and hence the correct interference pattern), we must square those amplitudes or complex number weightings[1].

The fundamental picture of a particle in quantum mechanics is that all alternative possiblities open to the system co-exist in a 'pure state' rather than a mixture of states until the instant when we observe it. It is important to get this distinction between a 'pure' and a 'mixed' state. A pure state is indivisible, it is a unity which we can represent as a 'superposition' of all the possible alternatives. The mixed state, however, is a mixture where the different states *really* exist in different proportions. The act of observation seems to put an end to this almost dream-like pure state into one of the possibilities that previously existed only as a potential. Hence, the observer seems to somehow determine the fate of the particle by 'collapsing' all its possibilities into a state of definiteness.

This is often told as the parable of Schrödinger's cat, kept in a box with a radioactive nuclide which might undergo radioactive decay and trigger a mechanism releasing cyanide gas to kill the cat. The quantum state of the cat therefore, is a superposition of being dead and being alive, until the instant when the observer opens the box, when it is either definitely dead, or definitely alive. Apart from being unkind to cats, this parable also suffers from human-observer chauvinism. One might raise the serious objection as to whether the cat, too, has a right to observe. Nevertheless, there is a school of thought which believes that it is the act of observation by the *human* consciousness which makes definite things happen.

We can generalize this picture to a particle, or a system that can exist in n possible states, then all of those possibilites are given a complex number weighting, the sum of all of which constitutes its quantum state, or wave function, ψ (Greek letter 'psi'). Classically, a particle has a definite position in space, and a definite momentum, which tells one where it is going to go next. Quantum mechanically, however, *every single position that the particle might have is available to it,* which together make up its wave function. A corresponding wave function can be composed of all the possible momenta it could have. The position and momentum wave functions are connected by a mathematical transformation called the *Fourier transform.* Position and momentum are related by a deep complementarity: a particle which has a precise position in space is completely indeterminate in its momentum; conversely, one which has a definite momentum will be completely delocalized in space, as can be seen in the Fourier transform diagrams (Fig. 10.3). A particle with definite position has a position wave function which is very sharply peaked at the value of x in question, all the amplitudes being zero elsewhere. Such a position wave function is a *delta function*. The momentum wave function corresponding to it is completely spread out

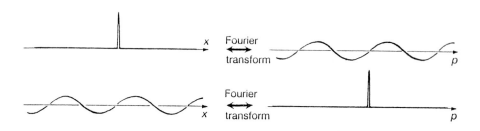

Figure 10.3 Fourier transforms between position (x) and momentum (p) (see text)[2].

in momentum values. Conversely, a particle with a delta function in momentum will be completely delocalized in space. Another way to express this is in terms of Heisenberg's *uncertainty principle*, which states that position and momentum of a particle cannot both be known precisely,

$$\Delta x \, \Delta p \geq h/4\pi$$

where Δ means 'uncertainty of' x and p are position and momentum respectively, and h is Planck's constant. This means that a measurement which gives a precise position value will be completely uncertain with respect to momentum, and *vice versa*, one giving a precise momentum value will be completely uncertain with respect to position. An analogous relationship for time, t, and energy, E, exists,

$$\Delta E \, \Delta t \geq h/4\pi$$

although, as we shall see in Chapter 11, this relationship is really quite different from the position/momentum uncertainty. The latter refers to the quantitites measured *at the same instant of time.* The energy/time relationship refers instead to the difference in energy which can be measured to any degree of accuracy at two different instants, and hence, is not the uncertainty in energy at a given instant.

The EPR Paradox and Quantum Entanglement

No description of quantum mechanics at any level is complete without the Einstein, Podolsky and Rosen paradox (the EPR paradox for short). Although Einstein contributed substantially to the development of quantum theory, he could never accept the description of reality in terms of the wave function consisting of a combination of complex amplitudes. This indefiniteness bothered him as something told him that, "God does not play dice!". He assumed there must be a deeper structure beneath quantum mechanics which presents reality without the contradictions inherent in the accepted theory. One person who continued in this line of thinking is David Bohm (see Chapter 12).

In order to try to show that a deeper, definite structure of reality exists. Einstein, Podolsky and Rosen proposed a thought experiment, or *gedanken* experiment (which in the end overcame their objections, especially when the experiment was eventually, actually carried out). The following is a later version due to Bohm[3]. Suppose two particles which are prepared in pairs - say an electron and a positron - with complementary properties such as spins up and down, move apart in opposite directions. By the conservation of momentum, the spins must add up to zero, because that is the spin of the initial central particle from which the pair has been generated. This has the implication, according to quantum theory, that if we measure the spin of the electron in whatever direction we choose, the positron must spin in the opposite direction, and *vice versa*. The two particles could be miles or light-years apart, yet the very choice of measurement on one particle seems instantaneously to fix the axis of spin of the other.

This remarkable result - which has been predicted and subsequently confirmed by experiment - cannot be explained by assuming that there are pre-set probabilities of the electron and positron pair to be in different combinations of spins, or that the two measuring machinery produces ordinary probabilistic responses, the probabilities being determined by the machinery.

One can increase the number of possible alternative settings for different directions of spins in the measurement process, say, in addition to spin 'up' or 'down', also 'left' or right' and other angles in between, but that does not alter the findings:

If the settings on both sides are the same, the results of the two sides always disagree. If the settings are spun at random, completely independently of each other even after the two particles have separated, then the two measurements are as equally likely to agree as to disagree.

This combination of results is incompatible with any local interaction model as the signals cannot propagate faster than light, and there is no set of prepared answers which can produce the quantum mechanical probabilities. The obvious conclusion that has to be drawn is that the effect of measurement (or 'collapse of the wave function') of one particle is instantaneously communicated to the other one. It is as though they are still one single coherent system, or in a pure state (see above), much like the single photon that passes through two slits at once, or the neutron that interferes with itself. A coherent system has neither space nor time separation, so the 'collapse' of one part is 'instantaneously' communicated to the other part, regardless of how great a distance exists between the two parts. By extrapolation of the experimental results, it must mean that the two parts could be light years apart, and still the 'collapse' of the wave function of one particle instantaneously collapses that of the other as well.

The concept of 'entanglement' was introduced by Schrödinger in 1935 to describe this phenomenon as part of his formal discussion of the EPR paper proposing the experiment. The two particles are, as it were, entangled with each other in a pure or coherent state[4]. It turns out that the two particles do not even have to be prepared together so that they are originally one system. Experimentally, one can even allow any two particles, neutrons, electrons, or photons, to be produced at distant and unrelated sources. As soon as they have come together and interacted, they become entangled with each other long after they have collided and separated. They have become one[5] coherent quantum system. This result is very significant, and I shall come back to it in the last chapter.

Quantum Coherence

The previous experiments bring out the important characteristics of the pure or coherent quantum state: delocalization, inseparability, nonlocal interactions. In order to to examine further what quantum coherence entails, let us return to the two-slit experiment as depicted in Figure 10.1. Recall that when both slits are opened, even single photons - generated one at a time at very low light intensities - behave as waves in that they seem to pass through both slits at once, and, falling upon the photographic plate, produces a pattern which indicates that each photon, in effect, interferes with itself! The intensity or brightness of the pattern at each point depends on a 'probability' that light falls on to that point.

The 'probability' is placed between quotation marks because, as said before, it is not probability in the ordinary sense. One way of representing these special probabilities is as correlation functions consisting of the product of two complex amplitudes. Light arriving at the point p on the photographic plate (Fig. 10. 1) has taken different paths, tp and bp. The intensity at p is then given as the sum of four such correlation functions:

$$I = G(t,t) + G\,(b,b) + G(t,b) + G\,(b,t)$$

where $G(t,t)$ is the intensity with only the top slit opened, $G(b,b)$ the intensity with only the bottom slit opened, and $G(t,b)+G(b,t) = 2G(t,b)$ is the additional intensity (which take on both positive and negative values) when both slits are opened. At different points on the photographic plate, the intensity is

$$I = G(t,t) + G(b,b) + 2\,|\,G(t,b)\,|\cos\theta$$

where θ is the angle of the phase difference between the two light waves.

The fringe contrast in the interference pattern depends on the magnitude of $G(t,b)$. If this correlation function vanishes, it means that the light beams coming out of t and b are uncorrelated; and if there is no correlation, we say that the light at t and b are incoherent. On the other hand, increase in coherence results in an increase in fringe contrast, i.e., the brightness of the bands. Since $\cos\theta$ is never greater than one (i.e., when the two beams are perfectly in phase), then the fringe contrast is maximized by making $G(t,b)$ as large as possible and that signifies maximum coherence. But there is an upper bound to how large $G(t,b)$ can be. It is given by the Schwarz inequality:

$$G(t,t,)G(b,b) \geq |\,G(t,b)\,|^2$$

The maximum of $G(t,b)$ is obviously obtained when the two sides are equal:

$$G(t,t)G(b,b) = |\,G(t,b)\,|^2$$

Now, it is this equation that gives us a description of quantum coherence. A field is coherent at two space-time points, say, t and b, if the above equation is true. Furthermore, we have a coherent field if this equality holds for all space-time points, X_1 and X_2. This coherence is called first-order coherence because its refers to correlation between two space-time points, and we write it more generally as,

$$G_{(1)}(X_1, X_1)G_{(1)}(X_2, X_2) = |\,G_{(1)}(X_1, X_2)\,|^2$$

The above equation tells us that, paradoxically, the correlation between two space-time points in a coherent field *factorizes*, or decomposes neatly into the self-correlations at the two points separately, and that this decomposability or factorizability is a sufficient condition for coherence. It is important to remember that the coherent state is a pure state and *not* a mixture of states (see previous sections). Factorizability does not mean that it can be factorized into a mixture. But it does lead to something quite unusual. What it means is that any two points in a coherent field will behave statistically independently of each other. If we put two photon detectors in this field, they will register photons independently of each other.

Coherence can be generalized to arbitrarily higher orders, say, to n approaching ∞, in which case, we shall be talking about a fully coherent field. If nth order coherence holds, then all of the correlation functions which represent joint counting rates for m-fold coincidence experiments (where $m<n$) factorize as the product of the self-correlations at the individual space-time points. In other words, if we put n different counters in the field, they will each record photons in a way which is statistically independent of all the others with no special tendency towards coincidences, or correlations (see Glauber[6]). Coherence can therefore exist to different orders or degrees[7].

Quantum Coherence and Living Organization

A coherent state thus maximizes both global cohesion and also local freedom! Nature presents us a deep riddle that compels us to accommodate seemingly polar opposites. What she is telling us is that coherence does not mean uniformity: where everybody must be doing the same thing all the time. An intuitive way to think about it is in terms of a symphony orchestra or a grand ballet, or better yet, a jazz band where every individual is doing his or her own thing, but is yet in tune or in step with the whole. This is precisely the biochemical picture we now have of the living system: micro-compartments and microdomains, right down to molecular machines, all functioning autonomously, doing very different things at different rates, generating flow patterns and cycles of different spatial extensions, yet all coupled together, in step with one another and hence, with the whole organism.

Many of the most paradoxical properties of the living system follow from coherence defined in this more rigorous sense. For example, factorizability optimizes communication by providing an uncorrelated network of space-time points which can be modulated instantaneously by specific signals. Furthermore, it provides the highest possible fringe contrast (or visibility) for pattern recognition, which may account for the great specificities in the response of organisms to diverse stimuli. The factorizability of coherent fields may also underlie the efficiency of bioenergetic processes in two respects. First, it concentrates the highest amount of energy of the field in localized zones by constructive interference, as well as creating effectively field-free zones within the field by destructive interference. Second, since the higher order correlations are the lowest in a completely coherent field, the smallest possible amount of energy is subject to correlated transfer between an arbitrarily large number of space-time points in the field with minimum loss. A coherent field is also fluctuationless or noiseless, the sort that any communications engineer working in radio-frequencies, for example, would say is coherent[8].

One way to be coherent is to occupy a single mode, as in superconductivity and superfluidity. That 'mode' does not have to be a single frequency, however, it is only necessary for it to represent one degree of freedom. Hence, there can be a broad band of frequencies coupled together or inter-communicating, so that energy fed into any frequency can be propagated to all other frequencies, as has been observed for light emission in living systems (see previous chapter).

I have argued in earlier chapters that quantum coherence is characteristic of living systems for two reasons. First, as molecular machines are *quantum* machines, quantum coherence must necessarily be involved in their coordination. Second, the rapidity of long range coordination in the living system is such as to rule out classical coherence due to classical phase transitions as exemplified by the Bénard convection cells. However, time is a subtle (and difficult) concept, as pointed out in connection with the space-time structure of living organisms.

Consider the consequence of coherence on energy storage in the living system. Coherence is associated with a time and a volume over which phase correlation is maintained. The coherence time for a quantum molecular process is just the characteristic time interval τ over which energy remains stored in McClare's formulation of the second law given in Chapter 3. So, in conformity with the second law of thermodynamics, the longer the coherence time, τ, the more extended is the timescale over which efficient energy transfer processes can take place, provided that the relaxation times are either much less than τ, or, in the quasi-equilibrium approximation, if they take place slowly with respect to τ. In other words, *efficient energy transfer processes can in principle occur over a wide range of timescales, depending on the coherence time-structure in the system*[9].

The existence of time (as well as space) structure in living systems also has its own interesting consequences of which I can only give a hint here. Time structure manifests itself most clearly in the range of biological rhythms that extend over some ten orders of magnitude from the millisecond oscillations of membrane action potentials to 10^7s for circannual rhythms, which are coherent over varying spatial domains from single cells to entire organs and from whole organisms to populations of organisms. A coherent space-time structure theoretically enables 'instantaneous' communication to occur over a range of time scales and spatial extents. What this implies in practice is a vast unexplored area, as the notion of nonlinear, structured time this entails is alien to the conventional, western scientific framework that this book is largely based upon. We shall go into the problem of time in more detail in the final chapter.

We may now offer a tentative answer to at least part of a question which was posed at the beginning of this book: what is it that constitutes a whole or an individual? It is *a domain of coherent, autonomous activity*. Defined thus, it opens the way to envisaging individuals which are aggregates of individuals, as, for example, a population or a society engaging in coherent activities[9]. As coherence maximizes both local freedom and global cohesion, it defines a relationship between the individual and the collective which has previously been deemed contradictory or impossible. The 'inevitable' conflict between the individual and the collective, between private and public interests, has been the starting point for all social as well as biological theories of western society. Coherence tells us it is not so 'inevitable' after all. Eminent sociologists have been deploring the lack of progress in sociology, and saying that it is time to frame new questions. Perhaps sociology needs a new set of premisses altogether[10]. In a coherent society, such conflicts do not exist. The problem is how to arrive at such an ideal state of organization that in a real sense, nurtures diversity (and individuality) with universal love[11].

Let us recapitulate the main results of the enquiry thus far into Schrödinger's question of 'what is life?', or 'can life be accounted for by physics and chemistry?'. I have shown how the intricate space-time structure of the living system cannot be accommodated within the statistical nature of the laws of thermodynamics, which therefore, cannot be applied to the living system without some kind of reformulation. This space-time structure arises at least partly as the consequence of energy flow, and is strongly reminiscent of the 'dissipative structures' or non-equilibrium phase transitions that can take place in physicochemical systems far from thermodynamic equilibrium. Energy flow organizes and structures the system in such a way as to reinforce the energy flow. Thermodynamics and its translation into the mechanical properties of molecules in statistical mechanics apply conventionally to systems which are unorganized, i.e., devoid of any space-time structure. What is required for our understanding of living organisms is not so much nonequilibrium as opposed to equilibrium thermodynamics - for the living system may possess isothermal equilibrium machines as well as far from equilibrium and quantum molecular machines - but perhaps a *thermodynamics of organized complexity*. I confess to have only a vague idea of what that entails, but it would use the concept of *stored* energy rather than *free* energy, as the former can be precisely defined in terms of the spatial extent and temporal duration of storage.

Another reason why conventional thermodynamics does not provide a sufficient explanation of the living system is because thermodynamics has its origins in describing the transformation of heat energy into mechanical work. The predominant energy transductions in the living system are instead, electronic, electric and electromagnetic, as consistent with the primary energy source on which life depends as well as the electromagnetic nature of all molecular and intermolecular forces. Furthermore, given the organized, condensed state of the living system, it is predicted that the most general conditions of energy pumping would result in a phase transition to a dynamically coherent regime where the whole spectrum of molecular energies can be mobilized for structuring the system, and for all the vital processes which make the system alive. Organisms are coherent space-time structures maintained far from thermodynamic equilibrium by energy flow. This enables them to store and mobilize energy with characteristic rapidity and efficiency. It also has profound implications on the nature of knowledge and knowledge acquisition, as well as issues of determinism and freewill, which are dealt with in the final chapters.

I make no claims to having solved the problem of life. I have sketched with a very broad brush where some of the clues may lie, pointing to a number of new and promising areas of investigations on bioelectrodynamics and nonlinear optical properties of living organisms. In the process, I have also raised many more subsidiary questions which I hope will keep the 'big' question alive.

Notes

1. For a readable and authoritative account see Penrose, R. (1989), Chapter 6.

2. I thank Oliver Penrose for pointing this out to me.

3. See Li (1992) for a formal argument which throws further light on this idea.

4. See Zajonc (1991).

5. See Glauber (1969).

6. This coherent quantum state description based on the quantized electromagnetic field approaches classical description when the system is coherent, see Goldin (1982) Chapter 8.

7. See Ho and Popp (1993).

8. See Ho (1993) for a more detailed discussion on these points.

9. I have discussed the possibilities for a 'coherent society' in a recent paper. See Ho (1992a).

10. See Ho (1992b).

CHAPTER ELEVEN

THE IGNORANCE OF THE EXTERNAL OBSERVER

The Meaning of Life and All That

Time has come for us to consider some of the implications of all that has gone before. Any answer, however tentative it may be, to the question of "what is life?" naturally has something to say on the meaning of life, as I intimated right at the beginning. This chapter is really about the issue of consciousness, which is also that of knowledge and of our relationship to the known, which ultimately encompasses all of nature.

Let us begin by looking more closely at the relationships between entropy and the concepts of knowledge, information and order in the context of the living system. These relationships have exercised western scientists and philosophers from the very first, and have remained the subject of lively debate to the present day. I believe that our new understanding of the living system may have something to offer towards the resolution of some of the central issues.

Is Entropy Subjective?

The concepts of thermodynamics took shape during the industrial revolution when attempts were made to determine how much mechanical work can be made available from steam engines. In the process, it was noticed that not all of the heat energy supplied can be transformed into work - some of it invariably gets lost, or becomes dissipated. This dissipated energy goes to make up entropy. Statistical mechanics interprets entropy in terms of the number of microstates the system can have, but it remains a mere analogue of the thermodynamic entity. Many irreversible processes have a well-defined and measurable macroscopic entropy change, and yet cannot be consistently represented in statistical mechanical terms.

One of these situations, described by Gibbs, concerns the entropy of mixing of perfect gases, and has become known as 'Gibbs paradox'. The entropy of mixing generated by, say, removing a partition separating the two gases occupying equal volumes on either side of the divide is always the same, and does not depend on how similar or otherwise the gases are. However, if the gases were *identical*, then the entropy of mixing has to be zero. But what if the two samples of the gas, previously thought to be identical are then found to be different after all? We must now say there is an increase in entropy to their mixing where previously we said it was zero. Maxwell writes,

"It follows from this that the idea of dissipation of energy depends on the extent of our knowledge....Dissipated energy is energy which we cannot lay hold of and direct at pleasure, such as the energy of the confused agitation which we call heat. Now confusion, like the correlative term order, is not a property of things in themselves, but only in relation to the mind which perceives them."[1]

Maxwell's idea - that entropy is a subjective concept which measures our ignorance - is all of a piece with the marked tendency among scientists in the 19th Century to regard any manifestation of randomness in natural phenomena as simply due to a lack of sufficient knowledge on our part, rather than as something inherent in nature[2]. Although Maxwell emphasized the statistical nature of the second law, he believed it was merely our knowledge of the world that is statistical, and not the world itself. Similarly, Gibbs, commenting on the probabilitistic foundations of statistical mechanics, said that probability refers to something which is 'imperfectly known'. And G.N. Lewis writes in 1930,

"The increase in entropy comes when a known distribution goes over into an unknown distribution ....Gain in entropy always means loss of information, and nothing more. It is a subjective concept, but we can express it in its least subjective form, as follows. If, on a page, we read the description of a physicochemical system, together with certain data which help to specify the system, the entropy of the system is determined by these specifications. If any of the essential data are erased, the entropy becomes greater; if any essential data are added the entropy becomes less."[3]

The same general viewpoint has been adopted by many scientists, the belief that entropy is a measure of the incompleteness of knowledge is one of the foundations of modern information theory, as we shall see.

Denbigh and Denbigh comment that although the use of less information than is available in certain aspects of statistical mechanics does result in a greater calculated entropy, *more* information can also lead to an increase in entropy. For example, information is gained when a spectral line of a substance, previously believed to be a singlet - one corresponding to a single electronic transition - is actually found to be a multiplet corresponding to transitions involving multiple electronic levels. This gain in information, however, will require the revision of the spectroscopic entropy upwards, as it implies the existence of previously undetected microstates accessible to the system.

Furthermore, entropy cannot be entirely subjective because it is an observable quantity, and its value can be determined by the observable data such as energy, volume and composition. And in that sense, entropy is no more subjective than any physical property, such as density, or vapour pressure, which is a function of all variables determining the state of the system.

However, entropy may not be a *property* of a system as such, for its absolute magnitude cannot be determined directly - even though it is supposed that all substances have zero entropy at absolute zero temperature (see Chapter 5). For one thing, we do not know anything about the entropy within the nucleus of the atom. In practice, we can only measure a *change* in entropy between two states along a reversible path. Thus, it should be called a *dispositional* property - one which can be observed only under certain circumstances, or when certain operations are carried out.

My own sympathy lies with the Denbighs especially with regard to entropy being a dispositional property. However, it is a dispositional property, not just of the system, but also of the processes that take place in it. The dispositional nature of entropy can take on a strong form in that it is *generated* only under certain circumstances or when the system is prepared in certain ways, as opposed to a weaker form where it may be generated under all or most circumstances but only observed in some of them.

Of course, as active agents who can set up experiments and choose what to measure or observe, we can influence the generation of entropy as well, so one might say that entropy is also a measure of our inability to influence the system. This demonstrates that in a very real sense, we participate in defining a process of measurement in partnership with nature, and out of this act, properties emerge which are neither those of things in themselves nor pure mental constructs, but an inextricable entanglement of both. This is a generalization of the inseparability of observer and observed that is at the heart of quantum measurement.

Thus, the subjectivist-objectivist dichotomy is falsely drawn. Subjectivity is an anthropomorphic, anthropocentric concept, and suffers from the same human chauvinism that attributes to the *human* observer alone the power to make definite things happen in the act of quantum mechanical measurement (see previous chapter). After all, other natural entities are entitled to their properties and propensities which are not just subject to our arbitrary intellectual whim and dictate. On the contrary, they require our active participation in the process of measurement in order to make manifest those properties which become part of our knowledge[4], the other part being the theory which relates those properties to one another[5].

To return to the dispositional nature of entropy, it is now becoming increasingly evident that many of the energy transducing processes in the living system may *generate* no entropy at all. And, if the living system does represent a coherent regime with, in effect, a single degree of freedom, then its entropy must certainly be close to zero. This has important implications for knowledge and knowledge acquisition. Before we come to that, let us deal more precisely with the measurement of entropy and of information.

Entropy and information

Entropy, as we have seen, may be defined in terms of the number of possible microstates in a closed system with a given volume, chemical composition and energy:

$$S_{BP} = k \ln W \qquad (1)$$

where S_{BP} is Boltzmann-Planck's entropy, as they were the first to use this analogue. The microstates are actually all the quantum states of the *macroscopic* system defined in terms of the wave functions of the coordinates of the particles in the system[6]. An astronomical number of quantum states are always comprised within a given thermodynamic system: a single gram of hydrogen already contains 10^{23} molecules! In the Boltzmann-Planck formulation, all the quantum states have the same energy and are by hypothesis, equally probable, i.e.,

$$P_i = 1/W \qquad \Sigma_i P_i = 1$$

where P_i is the probability of the ith quantum state and all of the P_i sum to 1.

In order to allow for fluctuations in energies, Gibbs put forward an alternative formulation for a system at thermal equilibrium with a large reservoir at a given temperature where there are now quantum states with different quantized energies, $e_1, e_2, ... e_j$. The probability of the jth energy state (or energy *eigenstate* as it is technically called), is then,

$$p_j = \frac{exp\ (-e_j\ /kT)}{\Sigma_j\ exp\ (-e_j\ /kT)} \qquad \Sigma_j\ p_j = 1$$

where k is Boltzmann's constant and T the temperature. Entropy now takes the form,

$$S_G = -k\ \Sigma_j\ p_j\ \ln\ p_j \qquad (2)$$

where S_G is Gibbs entropy which sums over the j energy states accessible to the system. Equation (2) reduces to equation (1) when all of the p_j s are equal to $1/W$. Both entropy measures become zero in the pure or coherent state, and take on maximum values at equilibrium, which is the maximally mixed state.

In 1949, Claude Shannon put forward an expression for the amount of information which is contained in messages sent along a transmission line. Suppose a message is to be composed of a string of four symbols which can only take one of two alternatives, 0 or 1, e.g., 0101, 0011, 1101, etc., then there are $2^4 = 16$ possible messages. If the zeros and ones are equally probable, then the 'information' contained in each message can be defined as being $\log_2 2^4 = 4$ 'bits'. And the information per symbol is hence one bit. The sequence of 4 symbols actually involves a succession of binary choices. If there are 4 alternative symbols instead of two, then each decision would have involved a choice of one out of 4, and any sequence of four symbols would have the information,

$$\log_2 4^4 = \log_2 2^8 = 8 \text{ bits}$$

so the information per symbol being now 2 bits. Of course, the number of alternative symbols may vary, and they may not be equally probable. An example is the English alphabet, which has 26 alternative symbols of widely different probabilities of occurrence. In general, a simple logarithmic formula could be written down for the measure of information or 'uncertainty' per symbol:

$$H = -\Sigma^n p_i \log_2 p_i \qquad (3)$$

where n is the number of alternative symbols and p_i is the probability of the ith symbol. Now, the similarity between equations (2) and (3) is such that it is indeed tempting to equate information with entropy, as Shannon has done.

Right from the beginning however, there were difficulties with both concepts, let alone the legitimacy of equating one with the other. Whose information is it anyway? The transmitter's or the receiver's? And where is the entropy? Is it located in the physical communication system which is after all indifferent to the actual messages sent?[7]

Information, Entropy, Order and Knowledge

A possibly more rigorous application of the concepts of information and entropy may be in the process of knowledge acquisition. The similarity of the H function and the S function suggests that the entropy is directly proportional to the information we would have if we know which microstate the system is in.

Thus, when we say that thermodynamic equilibrium is a state of maximum entropy, we are at the same time asserting that we are maximally ignorant about which microstate it is in - as it has equal probability of being in any one of the astronomical number of them. As Brillouin writes,

"Entropy is usually described as measuring the amount of disorder in a physical system. A more precise statement is that entropy measures lack of information about the actual structure of the system. The lack of information introduces the possibility of a great variety of microscopically distinct structures, which we are, in practice, unable to distinguish one from another."[8]

In this passage, Brillouin on the one hand acknowledges and asserts that the system subject to measurement has a structure of its own, and on the other hand, adopts a completely subjectivist view of entropy. So, who has the entropy? The observer or the observed? Morowitz[9] offers the following examples to illustrate more concretely the relativity of work and entropy, or rather, their dispositional character according to the information available.

Let us first examine a situation where a single molecule of a perfect gas is in one of two adjacent chambers connected by a valve, and the whole is immersed in a large reservoir at temperature T (see Fig. 11.1).

On opening the valve, a change in entropy will occur. We recall that Boltzmann's entropy is, $S = k \ln W$, where W is the number of possible microstates of the system. The change in accessible microstates will be proportional to the change in volume, hence,

$$\Delta S = k \ln V_2/V_1$$
$$= k \ln 2$$

If an observer knew which chamber contains the molecule, then she would have one bit of information, which is the amount involved in a binary choice. Moreover, this knowledge would enable her to rig up the valve to a piston and get work out of it. As this is an 'expansion' occurring isothermally, the total energy - which is a function of temperature only - does not change, i.e.,

$$\Delta U = T\Delta S - \Delta W = 0$$
$$\Delta W = T\Delta S$$
$$= kT \ln 2$$

If the information, as to which chamber the molecule is in, could be obtained without doing work, we would have a violation of the second law. So one must conclude that in order to obtain one bit of information, one has to do at least $kT \ln 2$ units of work, or alternatively, cause an increase in $k \ln 2$ units of entropy somewhere in the system. This is Brillouin's general conclusion on the relationships between information, entropy and work.

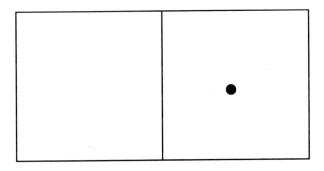

T

Figure 11.1 A single molecule of a perfect gas in one of two chambers connected by a valve and at equilibrium at temperature T.

Consider now the same two adjacent chambers of equal volume immersed in a large reservoir at temperature T, but one of the chambers contains a mole of a perfect gas i.e., N_0 molecules, where N_0 is Avagadro's number, while the other is empty (see Fig. 11. 2, top left diagram).

When a connecting valve is opened between the two chambers, an increase in entropy results:

$$\Delta S = N_0 k \ln (V_2/V_1) = R \ln 2$$

where $R = N_0 k$ is the gas constant. As before, the maximum amount of work which can be extracted from the system is,

$$\Delta W = RT \ln 2$$

This amount of work is the same even if we increase the number of chambers to 4, 8, 16, 32 or more, so long as only half of the chambers are filled with the mole of gas initially. If we carry this argument to molecular dimensions, we will have a distribution of gas hardly different from the completely random distribution that we would have obtained if the gas were originally at thermodynamic equilibrium within the entire volume. But the amount of work obtained in a reversible isothermal expansion is still $RT \ln 2$, just as the entropy change is still $R\ln2$; whereas in an equilibrium system both measures would be zero. This is paradoxical, as entropy is supposed to be a state function. We now have two systems in the same state (at equilibrium) but with different entropy measures depending on how the systems arrived at their final states.

The resolution to this paradox is that in the first case, we assumed that the work available is $RT \ln 2$. But in order to get the actual work, we have to know which chamber is filled and which empty. This is a binary decision requiring $RT \ln 2$ work units per molecule, as shown above. For a system with 1 mole or N_0 molecules, the work required would be $N_0 kT \ln 2$ or $RT \ln 2$. Now, an experimenter who has prepared the system, and therefore knows which chambers contain gas and which do not, will be able to get the maximum work of $RT\ln2$ out of the system. However, for one who does not know, at least the same amount of energy would have to be expended in order to get the information, and hence no net work can be obtained from the system.

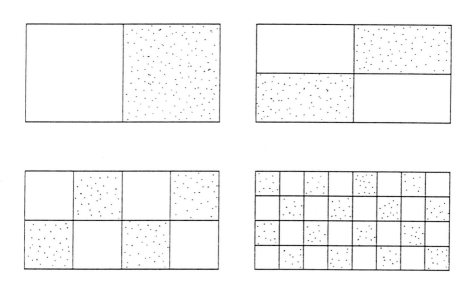

Figure 11.2 One mole of a perfect gas in different number of chambers at equilibrium at temperature T^{10}.

This appears to be a no-win situation as far as getting knowledge of the system is concerned. The system may have a very well defined coherent structure, which means that its real entropy is very low, as the number of microstates that the system accesses may be close to one. Yet the ignorance of the external observer seems to necessitate an exhaustive series of tests for the occupancy of each and every microstate that the system could exist in. As if that were not bad enough, if we were to actually carry out all those tests, the disturbance to the system would be such that we would no longer have the same system that we started with. Elsasser[11] states the problem as follows:

"If we try to predict with precision the future behavior of an organism we must, according to physics, know the initial values of the positions and velocities of its constituent particles with sufficient accuracy...[Otherwise] prediction of the future behavior of the system becomes correspondingly limited. Now, even in a relatively simple organism, a cell say, there are billions of orbital electrons which maintain a most intricate system of chemical bonds. The vital activities of the cell, especially metabolism, consists exactly in the changing of bonds, in the transfer processes which these bonding electrons undergo as time goes on. But quantum mechanics tells us that in order to localize an electron to within a reasonable accuracy ... a physical interaction of the electron with the measuring probe (e.g., X rays) must take place which, in the average, transmits a certain amount of interaction energy to the electron...The energy conferred upon the tissue by virtue of the measurements becomes then so large that vast changes must take place: the system will be radically altered and disrupted and is no longer even approximately equal to the system which was at hand before the measurements were undertaken."

Ignorance of the External Observer

Elsasser's remark encapsulates the dilemma of the external observer who has to determine the structure (microstate) of a living organism without disturbing it out of all recognition.

However, the ignorance of the external observer is really of no consequence to the system itself, which is far from ignorant of its own microstate. The problem of a system knowing its own microstate is vastly different from an external observer knowing its microstate. As Morowitz remarks, "The system is the microstate and by virtue of being the microstate has a great deal more knowledge of the microstate than is ever possible for an outside observer.....This fact is implicitly used in the design of experiments for molecular biology. We do not, as implied by the Bohr-Elsasser argument, set up experiments to measure the position and momentum of all atoms in a living system; we rather use a different strategy of allowing the experimental living system to generate answers to questions that we pose by the experimental set-up..."[12]

Furthermore, a system such as the organism, which is far from equilibrium, *yields* a lot more information about its microstates because it possesses macroscopic parameters which can readily be measured. In Chapter 4, we saw how macroscopic organization can spontaneously arise as energy flows through open systems, so much so that they can undergo nonequilibrium phase transitions to macroscopically coherent states characterized by a few macroscopic parameters (see also Chapter 7).

From the previous chapter, we also know that a coherent system is totally transparent to itself as all parts of the system are in complete, instantaneous communication. So in a precise sense, it *knows* itself completely. The effective degree of freedom of the system is one, and hence the entropy is zero. This suggests that the entropy of the living system can be expressed in terms of its deviation from coherence, i.e., where the degree of freedom becomes greater than one. In practice, however, in order to measure its deviation from coherence, we have to take into account the living system's coherent space-time structure, a quality which cannot be represented by the instantaneous Newtonian time that all western scientific theories are still based upon. Elsewhere, I have suggested that time itself is generated by process, specifically by the incoherence of action[13] (see also Chapter 12). In that respect, time has the same sense and direction as entropy.

There is another important implication as regards knowledge acquisition. The dilemma of the absolutely ignorant external observer betrays the alienation from nature that the traditional scientific framework of the west entails, for western science is premissed on the separation of the observer as disembodied mind from an objective nature observed[14]. This is also the origin of the subjective-objective dichotomy, which, when pushed to its logical conclusion, comes up against the seemingly insurmountable difficulty that in order to have sufficient information about the system, one has in effect to destroy it.

But as pointed out above, one can infer a great deal about the microstate of the system by *allowing* the system to inform. In other words, we must find ways of *communicating* with the system itself, rather than interrogating it, or worse, testing it to destruction. This is the reason why sensitive, non-invasive techniques of investigation are essential for really getting to know the living system.

Ideally, we ought to be *one* with the system so that the observer and observed become mutually transparent or coherent. For in such a pure, coherent state, the entropy is zero; and hence uncertainty and ignorance are both at a minimum. Perhaps such a state of enlightenment is just what Plato refers to as being one with the Divine Mind; or, as the *taoists* of ancient China would say, being one with the *tao*, the creative principle that is responsible for all the multiplicity of things[5]. It involves a consciousness that is delocalized and entangled with all of nature, when the awareness of self is heightened precisely because self and other are simultaneously accessed. I believe this is the essence of aesthetic or mystical experience.

This manner of knowing - with one's entire being, rather than just the isolated intellect - is foreign to the scientific tradition of the west. But I have just demonstrated that it is the only authentic way of knowing, if we were to follow to logical conclusion the implications of the development of western scientific ideas since the beginning of the present century. We have come full circle to validating the participatory framework that is universal to all traditional indigenous knowledge systems the world over[15]. I find this very agreeable and quite exciting.

Notes

1. Cited in Denbigh and Denbigh (1985) p. 3.

2. The description on entropy as a subjective concept is largely based on the excellent summary in Chapter 1 of Denbigh and Denbigh (1985).

3. Cited in Denbigh and Denbigh (1985) p. 4.

4. This issue is discussed at some length in Ho (1993).

5. See Ho (1992c).

6. This account is based on Denbigh (1981).

7. See Denbigh and Denbigh (1985) p. 104.

8. See Brillouin (1962).

9. See Morowitz (1968) pp. 124-8.

10. After Morowitz (1968) p. 127.

11. Elsasser (1966), cited in Morowitz, 1968.

12. See Morowitz (1968) pp. 131-2.

13. See Ho (1993).

14. See Ho (1990a).

15. See Ho (1993)

CHAPTER TWELVE

TIME AND FREEWILL

There is No Glass Bead Game

In this final chapter, I shall on the one hand, give free rein to my imagination without apology, as I try to convey what the physics of organisms means to me personally, and on the other, demonstrate explicitly that there is no discontinuity between the so-called 'hard' sciences such as physics and chemistry and the 'soft' sciences such as psychology and philosophy. This chapter is not just the icing on the cake, but stems from a deep conviction that knowledge, especially participatory knowledge, is what we live by, and so, there should also be no discontinuity between knowledge and action. There can be no 'glass bead game' to knowledge[1], especially scientific knowledge, for it matters far too much.

Bergson's 'Pure Duration'

The title of this chapter is taken from the French philosopher, Henri Bergson's fascinating book[2], with which I have established a relationship for at least 15 years, before I finally understand the message in the poetry in a way that allows me to write about it.

The second law of thermodynamics defines a time's arrow in evolution[3] and in that sense, captures an aspect of experience which we know intuitively to be true. By contrast, the laws of microscopic physics are time reversible: one cannot derive a time's arrow from them, nor are they affected by a reversal of time[4]. There is another sense in which time (as well as space) in physical laws does not match up to our experience. Newtonian time, and for that matter, relativistic time and time in quantum theory, are all based on a homogeneous, linear progression - the time dimension is infinitely divisible, so that spatial reality may be chopped up into instantaneous slices of immobility, which are then strung together again with the 'time line'. Real processes, however, are not experienced as a succession of instantaneous time slices like the successive frames of a moving picture. Nor can reality be consistently represented in this manner.

The mismatch between the quality of authentic experience and the description of reality given in western science has long been the major source of tension between the scientists and the 'romantics'. But no one has written more vividly on the issue of time than Bergson, with so few who could really understand him. He invites us to step into the rich flowing stream of our consciousness to recover the authentic experience of reality for which we have substituted a flat literal simulacrum given in language, in particular, the language of science.

In the science of psychology, words which express our feelings - love and hate, joy and pain - emptied of their experiential content, are taken for the feelings themselves. They are then defined as individual psychic entities (or psychological states) each uniform for every occasion across all individuals, differing only in magnitude, or intensity. Should we connect our mind to our inner feelings, what we experience is not a quantitative increase in intensity of some psychological state but a succession of qualitative changes which "melt into and permeate one another" with no definite localizations or boundaries, each occupying the whole of our being within this span of feeling which Bergson refers to as 'pure duration'.

Pure duration is our intuitive experience of inner process, which is also inner time with its dynamic heterogeneous multiplicity of succession without separateness. Each moment is implicated in all other moments (c.f. the organism's space-time structure described in Chapter 3). Thus, Newtonian time, in which separate moments, mutually external to one another, are juxtaposed in linear progression, arises from our attempt to externalize pure duration - an indivisible heterogeneous quality - to an infinitely divisible homogeneous quantity. In effect, we have reduced time to Newtonian space, an equally homogeneous medium in which isolated objects, mutually opaque, confront one another in frozen immobility.

Bergson emphasizes the need for introspection in order to recover the quality of experience. He opposes an inner "succession without mutual externality" to an outer "mutual externality without succession". He distinguishes two different selves, one the external projection of the other, inner self, into its spatial or social representation. The inner self is reach "by deep introspection, which leads us to grasp our inner states as living things, constantly *becoming*."

"..But the moments at which we thus grasp ourselves are rare, and that is just why we are rarely free. The greater part of the time we live outside ourselves, hardly perceiving anything of ourselves but our own ghost, a colourless shadow which pure duration projects into homogeneous space. Hence our life unfolds in space rather than in time; we live for the external world rather than for ourselves; we speak rather than think; we 'are acted' rather than act ourselves.."5

This passage anticipates the sentiment of the existentialist writers such as Camus and Sartre. A similar sentiment pervades T.S. Eliot's poetry. The following lines evoke strong echoes of Bergson's projected being outside ourselves,

> We are the hollow men
> We are the stuffed men
> Leaning together
> Headpiece filled with straw. Alas!
> Our dried voices, when
> We whisper together
> Are quiet and meaningless
> As wind in dried grass
> Or rats' feet over broken glass
> In our dried cellar
>
> Shape without form, shade without colour,
> Paralysed force, gesture without motion;

Bergson's protests were directed against one of the most fundamental assumptions underlying modern western science. It claims to express the most concrete, common-sensible aspect of nature: that material objects have simple locations in space and time. Yet space and time are not symmetrical. A material object is supposed to have extension in space in such a way that dividing the space it occupies will divide the material accordingly. On the other hand, if the object lasts within a period of time, then it is assumed to exist equally in any portion of that period. In other words, dividing the time does nothing to the material because it is always assumed to be immobile. Hence the lapse of time is a mere accident, the material being indifferent to it. The world is simply fabricated of a succession of instaneous immobile configurations of matter (i.e., a succession of equilibria), each instant bearing no inherent reference to any other instant of time. How then, is it possible to link cause and effect? How are we justified to infer. from observations, the great 'laws of nature'? This is essentially the problem of induction raised by Hume[6]. The problem is created because we have mistaken the abstraction for reality - a case of the fallacy of misplaced concreteness.

Whitehead's 'Organism' and Bohm's 'Implicate Order'

In order to transcend the philosophical ruin left in the wake of mechanical materialism, Whitehead[6] attempts to return to a kind of native realism, not unlike the panpsychism or pananimism that is usually attributed to the so-called primitive mind by western anthropologists. He rejects the existence of inert objects or things with simple location in space and time. As all nature is process, there is only the progressive realization of natural occurrences. For mind, he substitutes a process of 'prehensive unification' - a kind of perception that may or may not involve human consciousness. The realization of an occurrence is "a gathering of things into a unity" which defines a *here* and a *now,* but the things gathered into a unity always refer to other places and other times. The totality of occurrences thus consists of a pattern of flows and influences now diverging from one locus, now converging towards another in such a way that "each volume of space, or each lapse of time includes in its essence aspects of all volumes of space, or of all lapses of time".

An organism, according to Whitehead, is "a locus of prehensive unification". This just corresponds to a field of coherent activities which is sensitive to the environment, in our current language. Whitehead asserts that the fundamental particles of physics such as protons and electrons are organisms, as much as, at the other extreme, entire planets such as the Earth, are also organisms. Nevertheless, he does recognize gradations of organisms and hence of consciousness[7]. Each organism, in the act of prehensive unification, enfolds the environment consisting of others into a unity residing in a 'self', while aspects of the self are communicated to others. The realization of 'self' and 'other' are thus completely intertwined. The individual is a distinctive enfoldment of its environment, so each individual is not only constituted of others in its environment, but also simultaneously delocalized over all individuals. The society is thus a community of individuals mutually delocalized and mutually implicated. Individuality is also relative, for an organism can be part of the life history of some larger, deeper, more complete entity.

An obvious situation where nested individualities occur is in a society or a community of species. Societies evolve slowly, its slowly-changing variables define parameters for the evolution of individuals within it. The picture of nested individuality and constitutive mutuality is also consistent with Lovelock's Gaia hypothesis[8]: that the earth is one cybernetic system maintained far from thermodynamic equilibrium in conditions suitable for life by the actions of the 'organisms' (both physical and biological) within it. Not only are individuals part of a larger organism, but the substance, the very essence of each individual is constitutive of every other. This is all of a piece with the concepts of nonlocality and quantum entanglement that we have described in Chapter 10, giving further substance to the participatory worldview.

In a very real sense, we participate not only in our knowledge of nature but in creating reality in partnership with all of nature. A truly participatory consciousness would perceive this coherent delocalization of self and other as constitutive of its own being (see previous chapter) which not only gives it authentic knowledge, but also enpowers it to act appropriately and coherently.

The moral feeling arises from this primary perception of the mutual entanglement of self and other, ultimately, of all life.

The mutual enfoldment and unfoldment of the 'implicate' and 'explicate', between organism and environment, is precisely David Bohm's account of the quantum universe[9]. In analogy to the hologram, the implicate order of an object is contained in an interference pattern of light distributed throughout space, in which it can be said to be enfolded. By an act of unfoldment, however, the original form of the object could once again be made explicate.

In the latest version of Bohm's theory, the universe is pictured as a continuous field with quantized values for energy, momentum and angular momentum. Such a field will manifest both as particles and as waves emanating and coverging on the regions where particles are detected. This field is organized and maintained by the 'superquantum potential' which is a function of the entire universe:

"..What we have here is a kind of universal process of constant creation and annihilation, determined through the superquantum potential so as to give a world of form and structure in which all manifest features are only relatively constant, recurrent and stable aspects of the whole..."[10]

Whitehead's organicist philosophy is in many ways a logical progression from the demise of mechanical materialism which began towards the end of the previous century. The rise of thermodynamics introduced a new kind of conservation law: that of energy in place of mass. Mass was no longer the pre-eminent permanent quality. Instead, the notion of energy became fundamental, especially after Einstein worked out the famous mass-energy equation, and nuclear fission in the atomic bomb proved him right, with devastating consequences. At the same time, Maxwell's theory of electromagnetism demanded that there should indeed be energy, in the form of electromagnetic fields pervading throughout all space, which is not immediately dependent on matter. Finally, the development of quantum theory reveals that even the atoms of solid matter are thought to be composed of vibrations which can radiate into space. Matter loses solidity more and more under the steady scrutiny of relentless rationality.

Meanwhile, the Newtonian picture of homogeneous absolute time and space gives way to relativity. Each inertial frame of reference (associated with its own observer or prehensive organism) must be considered as having a distinct space-time metric different from those of other inertial frames. The organism has no simple location in space-time. Moreover, an organism can alter its space-time by its own motion or activity. It is possible that some organisms will no longer 'endure' under changes of space-time. Thus, the organism's space-time metric, and perforce, its *internal* space-time, cannot be regarded as given, but arises out of its own activities. The organization of these activities is also its internal *space-time structure*, which is not a quantity but a quality. Bergson's 'pure duration' is a quality of the same cloth.

Space-time Structure and Quantum Theory

The nature of space and time is fundamental to our theory of reality. The mismatch between the Newtonian universe and our intuitive experience of reality hinges on space and time. In fact, all subsequent developments in western science may be seen as a struggle to reinstate our intuitive, indigenous notions of space and time, which deep within our soul, we feel to be more consonant with authentic experience. But there has only been limited success so far.

Einstein's theory of special relativity substitutes for absolute space and absolute time a four-dimensional space-time continuum which is different for each observer in its own inertial frame. Space and time have become symmetrical to each other, but they remain definite quantities. In the usual quantum theory, on the other hand, space coordinates lose definiteness in becoming complex statistical quantities (see Chapter 10), but time remains a simple parameter as in classical mechanics.

Another problem in connection with time already mentioned is that the laws of physics in both classical and quantum mechanics as well as in relativity, are time-symmetric, i.e., they do not distinguish between past and future. Yet real processes seem to have an 'arrow of time'. So time ought to be related to real processes. If so, it would have the quality that Bergson refers to as pure duration. In other words, it would have a structure. Recently, the German physicist, Wolfram Schommers[11], is taking up just this problem.

Like both Whitehead and Bergson, he argues for the primacy of process, and in an interesting reformulation of quantum theory, shows how time and space are tied to real process.

He begins from a consideration of Mach's principle, according to which, particles do not move relative to space, but to the centre of all the other masses in the universe. In other words, absolute space and time coordinates cannot be determined empirically. Any change in position of masses is not due to the interaction between coordinates and masses, but entirely between the masses. However, neither relativity nor quantum mechanics have incorporated Mach's principle in their formulation.

If one takes account of Mach's principle, space-time must be considered as an auxiliary element for the geometrical description of real processes. In other words, real processes are projected to space-time or '(r,t)-space' from perhaps a more fundamental space - that which represents reality more authentically in terms of the parameters of interactions, i.e., momentum and energy - the '(p,E)-space'. The two spaces are equivalent descriptions and are connected by the mathematical device of a Fourier transformation. (Intermediate spaces, (r,E) and (p,t), can also be formed which are similarly connected.) The result is that time, as much as space, becomes statistical quantities in (p,E)-space where momentum and energy are parameters and take on definite values. Conversely, momentum and energy are statistical quantities in (r,t)-space where space and time are definite parameters.

The wave function for space-time, $\psi(r,t)$ leads to probability distributions for *both* space and time in (p,E)-space. This means that processes consisting of matter interacting *generate* space-time structures. In other words, space-time structures are caused by action, and in the limiting case of a stationary (or equilibrium) process and a free particle (one that is not subject to any external influence), no time- and space-structures are defined, i.e., the wave function $\psi(r,t) = 0$.

In Schommer's scheme, energy and time representations are complementary, and for nonstationary processes, an uncertainty relationship exists between them *which is of the same form as that between position and momentum* in conventional quantum theory. The consequence is that both energy structure and the internal time structure are different for different systems when compared to an external reference time structure such as a clock. For nonstationary processes, there will be probability distributions for all the variables, p, r, E and t, which are *internal* to the system; whereas for stationary processes, there is no system-specific time definable and the energy is constant, so only position and momentum take on probability distributions.

Another consequence is that wave and particle are no longer in contradiction. They coexist simultaneously in this new representation. For a particle is a point with definite momentum and energy in (p,E)-space, but *simultaneously* a wave in the (r,t)-space. Conversely, a particle in the (r,t)-space is simultaneously a wave in the (p,E)-space, and so wave and particle co-exist in different spaces *independently* of the measurement process. This removes the observer paradox, or the 'collapse of the wave function' mentioned in Chapter 10. In Schommer's scheme, the wave function never collapses, it only becomes either a particle or wave in one space, and its complement in the other space. This has fundamental implications for the nature of consciousness as we shall see.

By the same token, a free particle cannot be observed in (r,t)-space, for by definition, it does not interact with any other system, i.e., there is no exchange of momentum and energy. An observation implies an interaction process so that the particle is no longer free. Thus, knowledge acquisition is always participatory: a partnership between the knower and the known. It is both subjective and objective, as we have already concluded in connection with considerations on entropy and information in the previous chapter.

Organisms as Coherent Space-time Structures

Schommer's reformulation of quantum theory is especially relevant for understanding the space-time structure of living systems. If organisms are coherent space-time structures, then certain consequences should flow from that. As stated in Chapter 10, an individual is simply a field of coherent activity. Defined in this way, we can readily appreciate the sort of nested individualities that Whitehead speaks of, but with the added insight that individualities are spatially and temporally fluid entities, in accordance to the extent of the coherence established. Thus, in long-range communication between cells and organisms, the entire community may become one when coherence is established and communication occurs without obstruction or delay. Within the coherence time, there is no space separation, i.e., the usual spatial neighbourhood relationship becomes irrelevant. Similarly, within the coherence volume, there is no time separation, hence 'instantaneous' communication can occur. (Feelings can indeed spread 'like wild fire', and people everywhere can get caught up simultaneously in a sudden fervour.)

Within the living system, coherence times and coherence volumes are themselves determined by the relaxation times and volumes of the processes involved. We may envisage biological rhythms as manifesting a hierarchy of coherence times that define the time frames of different processes. This fits with Bergson's concept of pure duration, which, in one sense, we may identify as the time taken for the completion of a process. A heartbeat requires a full cycle of contraction and relaxation before we recognize it as such - the duration of a heart-beat is about one second in external reference time. Similarly, neurobiologists have recently discovered an endogenous 40hz rhythm that is coherent over the entire brain (see Chapter 7). This may define the duration of primary perception. Within that duration, which we can regard as the coherence time in that level of the the nested hierarchy of time structure, processes coherent with it will generate no time at all. The coherence volume, similarly, is the extent of spatial coherence within which there is no time separation.

This representation of individuals as coherent space-time structures implies that space and time, in terms of separation and passage, are both generated, perhaps in proportion to the incoherencies of action. (Thus, a coherent sage may well be living in a truly timeless-spaceless state, which is beyond our comprehension. I believe some of us get glimpses of that in a particularly inspired moment, or during an aesthetic or religious experience.)

In ordinary perception, on the other hand, the organism interacts with the environmental object, a perturbation propagates within the organism and is registered or assimilated in its physiology (in other words, enfolded). This results in time generation. The greater the wave function $\psi(r,t)$ changes (see below), or the greater the mismatch between object and subject, the longer the time generated. Conversely, the more match or transparency between object and subject, the less.

From the foregoing discussion, one can see that a coherent society (mentioned in Chapter 10) must obviously be one where social space-time structure matches both natural astronomical space-time as well as individual private space-time. This has considerable relevance for Illich's idea of convivial scales of machinery as well as communities and institutions[12]. When the latter become too large, as Illich points out, they have a way of enslaving us instead of serving and supporting us.

Determinism and Freewill

At the end of his book, "What is Life?", Schrödinger has this to say on determinism and freewill,

".. let us see whether we cannot draw the correct, non-contradictory conclusion from the following two premises:

(i) My body functions as a pure mechanism according to the Laws of Nature.

(ii) Yet I know, by incontrovertible direct experience, that I am directing its motions, of which I forsee the effect, that may be fateful and all-important, in which case I feel and take full responsibility for them.

"The only possible inference from these two facts is, I think, that I - I in the widest meaning of the word, that is to say, every conscious mind that has ever said or felt 'I' - am the person, if any, who controls the 'motion of the atoms' according to the Laws of Nature..."

He continues,

"..Consciousness is never experienced in the plural, only in the singular..."[13]

This is just what the state of coherence entails: a multiciplicity which is singular, the 'self' is a domain of coherence, a pure state which permeates the whole of our consciousness, much as Bergson has described.

The positing of 'self' as a domain of coherent space-time structure implies the existence of active agents who are free. Freedom in this context means being true to 'self', in other words, being coherent. A free act is thus a coherent act. Of course not all acts are free, since one is seldom fully coherent. Yet the mere possibility of being unfree affirms the opposite, that freedom is real,

"..we are free when our acts spring from our whole personality, when they express it, when they have that indefinable resemblance to it which one sometimes finds between the artist and his work."[14]

But as the 'self' is also distributed - being implicated in a community of other entities (see p. 174) - to be true to 'self' does *not* imply acting against others. On the contrary, sustaining others sustains the self, so being true to others is also being true to self. It is only within a mechanistic Darwinian perspective that freedom becomes perverted into acts against others[15].

According to John Stuart Mill[16], to be free, "must mean to be conscious, before I have decided, that I am able to decide either way." So defenders of freewill claim that when we act freely, some other action would have been equally possible. Conversely, proponents of determinism assert that given certain antecedent conditions, only one resultant action was possible.

As Bergson remarks, the problem itself is posed on the mechanistic assumptions of immobility and mutual externality of events. This gives rise to two equally unacceptable alternatives: either that an immobile configuration of antecedents is supposed to 'determine' another immobile configuration of resultants, or that at any frozen instant, to be or not to be are equally likely choices for a consciousness that is external to itself, which immediately leads us back to Cartesian mind-matter dualism that makes us strangers to ourselves. In the reality of process, where the self is ever becoming, it does not pass like an automaton from one frozen instant to the next. Instead, the quality of experience permeates the whole being in a succession without separateness in "a self which lives and develops by means of its very hesitations, until the free action drops from it like an over-ripe fruit."[16]

One might represent consciousness as a wave function that evolves, constantly being transformed by experience as well as overt acts. The issue of quantum indeterminism is a very deep one, but the picture of a wave function - a pure state - consisting of a total interfusion of feelings each of which occupying the whole being - is very like what Bergson describes. Such a pure state is an indivisible unity and must be distinguished from a mixture of states (see Chapter 10). Thus, the overt act, or choice, does follow from the antecedent, but it cannot be predicted in advance. One can at best retrace the abstract 'steps' and represent the evolution of the consciousness as having followed a 'trajectory'. In truth, the so-called trajectory has been traced out by one's own actions, both overt and covert up to that point. When one reinstate the full quality of our consciousness, we can see that there can be no identical or repeatable states, which, when presented again at any time, will bring about identical resultant states.

I suggest that the 'wave function' that is consciousness never collapses (c.f. Schommers, p.177), but is always changing and always unique as it is 'coloured' by all the tones of our personality and experience. Each significant experience becomes entangled in our being, constituting a new wave function, in much the same way that particles of independent origins become entangled after they have interacted (see Chapter 10). The reason we experience our macroscopic world so differently from the microscopic world may be tied up with the space-time structure of reality. It could be that, were we the dimensions of elementary particles, we too would experience our world as being inhabited by self-possessed, coherent beings with a free-will, and we would be most surprised to find that human-sized observers are trying to predict which state we are going to 'collapse' into next. In the same way, an external observer of galactic dimensions would find our human behaviour completely 'quantum' and unpredictable.

Gibson, a chief exponent of a process ontology in perception, has this to say on consciousness,
"The stream of consciousness does not consist of an instantaneous present and a linear past receding into the distance; it is not a 'travelling razor's edge' dividing the past from the future. Perhaps the present has a certain duration. If so, it should be possible to find out when perceiving stops and remembering begins. But it has not been possible... A perception in fact, does not have an end. Perceiving goes on."[18]

Nature is ever-present to us, as we are to ourselves. This ever-present is structured, as we have seen. Our experience consists of the catenation of events of different durations, which propagates and reverberates in and around our being, constantly being registered and recreated. What constitutes memory of some event is the continuing present for the over-arching process of which the event is part[19].

The universe of coherent space-time structures is thus a nested hierarchy of individualities and communities which come into being through acts of prehensive unification. It is a truly participatory, creative universe. Just as the organism is ever-present to itself during its entire life history, and all other individualities are ever-present to it, the universe is ever-present to itself in the universal duration where creation never ceases by the convocation of individual acts, now surfacing from the energy substrate, now condensing to new patterns, now submerging to re-emerge in another guise.

Reality is thus a shimmering presence of infinite planes, a luminous labyrinth of the active now connecting 'past' and 'future', 'real' with 'ideal', where potential unfolds into actual and actual enfolds to further potential through the free action and intention of the organism. It is a sea awash with significations, dreams and desires. This reality we carry with us, an ever-present straining towards the future. The act is the cause, it is none other than the creation of meaning, the realization of the ideal and the consumation of desire.

Notes

1. See Ho (1993), referring to Hesse (1943; 1970). Hermann Hesse's famous last novel describes the life of Joseph Knecht, Magister Ludi, or the supreme master of the Glass Bead Game. The game is one of pure intellect directed at the synthesis of the spiritual and aesthetic abstractions in diverse disciplines of all ages, and is the prerogative and *raison d'etre* of an entire spiritual institution, Castalia. Isolated within its enclaves and unsullied by reality, the chosen elite undertook arduous scholastic studies, the sole purpose of which was to create ever more intricate themes and variations of the game. Castalia and the Glass Bead Game developed as antithesis to the philistine, superficial bourgeoise society, intent on its own pursuit of conventional, establishment values. In the end, however, Joseph Knecht turned his back on Castalia , disillusioned with a life consecrated exclusively to the mind, recognizing not only its utter futility, but also its inherent danger and irresponsibility.

2. Bergson (1916).

3. See Saunders and Ho (1976), and references therein.

4. For an excellent accessible survey of time in physics and philosophy, see Flood and Lockwood (1986).

5. Bergson (1916) p. 231.

6. See Whitehead (1925).

7. See Ho (1993).

8. See Lovelock (1979); (1988).

9. See Bohm (1987).

10. Bohm (1987) p. 43.

11. See Schommers (1989).

12. Illich (1973).

13. Schrödinger (1944) pp.86-8.

14. Bergson (1916) p. 172.

15. See Ho (1992b).

16. Mill (1878), p.580-583, cited in Bergson, 1916, p. 174.

17. Bergson (1916) p. 176.

18. Gibson (1966).

19. See Ho (1992c) .

REFERENCES

Adey, W.R. "Collective Properties of Cell Membranes." In *Resonance and Other Interactions of Electromagnetic Fields with Living Systems. Royal Swedish Academy of Sciences Symposium*, May 22, 1989.

Alberts, B., Bray, D., Lewis, J., Raff, M., Roberts K. and Watson, J.D. *Molecular Biology of The Cell*, Garland Publishing, Inc., New York, 1983.

Astumian, R.D., Chock, P.B., Tsong, T.Y. and Westerhof, H.V. "Effects of Oscillations and Energy-driven Fluctuations on the Dynamics of Enzyme Catalysis and Free-energy Transduction." *Physical Review A* **39** (1989): 6416-35.

Baba, S.A. "Regular Steps in Bending Cilia During the Effective Stroke." *Nature* **282**(1979): 717-72.

Batlogg, B. "Physical Properties of High-Tc Superconductors." *Physics Today* (June, 1991): 44-50.

Becker, R.O. *Cross Currents: The Promise of Electromedicine, the Perils of Electropollution.* Jeremy P. Tacher, Inc., Los Angeles, 1990.

Bergson, H. *Time and Free Will. An Essay on the Immediate Data of Consciousness* (F.L. Pogson, trans.), George Allen & Unwin, Ltd., New York, 1916.

Berridge, M.J., Rapp, P.E., and Treherne, J.E., Eds. *Cellular Oscillators, J. Exp. Biol.* **81**, Cambridge University Press, Cambridge, 1979.

Bohm, D. "Hidden Variables and the Implicate Order." In *Quantum Implications. Essays in Honour of David Bohm* (B.J. Hiley and F.D. Peat, Eds.), pp. 33-45, Routledge and Kegan Paul, New York, 1987.

Breithaupt, H. "Biological Rhythms and Communications." In *Electromagnetic Bio-Information, 2nd ed.* (F.A. Popp, R. Warnke, H.L. Konig and W. Peschka, Eds.), pp. 18-41, Urban & Schwarzenberg, Munchen, 1989.

Bridgman, P.W. *The Nature of Thermodynamics*, Harvard University Press, Cambridge, Mass, 1941.

Brillouin, L. *Science and Information Theory, 2nd ed.*, Academic Press, New York, 1962.

Chang, R. *Physical Chemistry with Applications to Biological Systems, 2nd ed.*, MacMillan Publishing Co., inc., New York, 1990.

Chu, B. *Molecular Forces, Based on the Baker Lecture of Peter J. W. Debye*, John Wiley & Sons, New York, 1967.

Clegg, J.S. "Properties and Metabolism of the Aqueous Cytoplasm and Its Boundaries." *Am J. Physiol.* **246** (1984): R133-51.

Cope, F.W. "A Review of the Applications of Solid State Physics Concepts to Biological Systems." *J. Biol. Phys.* **3** (1975): 1-41.

Davydov, A.S. "Solitons and Energy Transfer Along P{rotein Molecules." *J. Theor. Biol.* **66** (1977): 379-87.

Davydov, A.S. *Biology and Quantum Mechanics*, Pergamon Press, New York, 1982.

Davydov, A.S. *Excitons and Solitons in Molecular Systems.*, Academy of Sciences of the Ukrainian SSR Institute for Theoretical Physics Preprint ITP-85-22E, Kiev, 1985.

Denbigh, K.G. *The Thermodynamics of the Steady State*, Mathuen & Co., Ltd., New York, 1951.

Denbigh, K.G. *The Principles of Chemical Equilibrium, 4th ed.* Chapter 11, Cambridge University Press, Cambridge, 1981.

Denbigh, K.G. "Note on Entropy, Disorder and Disorganization." *Brit. J. Phil. Sci.* **40** (1989): 323-32.

Denbigh, K.G. and Denbigh, J.S. *Entropy in Relation to Incomplete Knowledge*, Cambridge University Press, Cambridge, 1985.

Duffield, N.G. "Global Stability of Condensation in the Continuum Limit for Fröhlich's Pumped Phonon System." *J. Phys. A: Math. Gen.* **21** (1988): 625-41.

187

Edmonds, D.T. "Larmor Precession as a Mechanism for the Detection of Static and Alternating Magnetic Fields." *Bioelectrochemistry and Bioenergetics* (1992) in press.

Ehrenberg, W. "Maxwell's Demon."*Sci. Am.* **217** (1967): 103-10.

Elsasser, W. *Atoms and Organism*, Princeton University Press, Princeton, New Jersey, 1966.

Ferrier, J., Ross, S.M., Kanehisa, J. and Aubin, J.E. "Osteoclasts and Osteoblasts Migrate in Opposite Directions in Response to a Constant Electric Field." *J. Cell Physiol.* **129** (1986): 283-8.

Feynman, R.P. *Q.E.D. The Strange Theory of Light and Matter*, Penguin Books, Harmondsworth, 1985.

Flood, F. and Lockwood, M. eds. *The Nature of Time*, Blackwell, Oxford, 1986.

Fox, S.W. *Selforganization*, Adenine Press, Guilderland, New York, 1986.

Fröhlich, H. "Long Range Coherence and Energy Storage in Biological Systems." *Int. J. Quantum Chem.* **2** (1968): 641-49.

Fröhlich, H. "The Biological Effects of Microwaves and Related Questions." *Adv. Electronics and Electron. Phys.* **53** (1980): 85-152.

Galle, M., Neurohr, R., Altman, G. and Nagl, W. "Biophoton Emission from Daphnia Magna: A Possible Factor in the Self-regulation of Swarming." *Experientia* **47** (1991): 457-60.

Gibson, J.J. *The Ecological Approach to Visual Perception*, MIT Press, Mass., 1966.

Glasstone, S. *A Textbook of Physical Chemistry*, 2nd ed., Macmillan, London, 1955.

Glauber, R.J. "Coherence and Quantum Detection." In *Quantum Optics* (R.J. Glauber, Ed.), Academic Press, New York, 1969.

Goldin, E. *Waves and Photons, An Introduction to Quatnum Optics*, John Wiley and Sons, New York, 1982.

Gray, C.M., Konig, P., Engel, A.K. and Singer, W. "Oscillatory Responses in Cat Visual Cortex Exhibit Inter-columnar Synchronization Which Reflects Global Stimulus Properties." *Nature* **338** (1989): 334-37.

Gross, M. and Haroche, S. "Superradiance: An Essay on the Theory of Collective Spontaneous Emission." *Physcis Reports* **93** (1982): 301-96.

Gurwitsch, A.G. "The Mitogenic Rays." *Bot. Gaz.* **80** (1925): 224-6.

Haken, H. *Synergetics*, Springer-Verlag, Berlin, 1977.

Haroche, S. and Kleppner, D. "Cavity Quantum Electrodynamics." *Physcis Today* **42** (1989): 24-30.

Hess, G. "The Glycolytic Oscillator." *J. Exp. Biol.* **81** (1979): 7-14.

Hesse, H. *Magister Ludi, The Glass Bead Game* (R. and C. Winston, trans.), Bantam Books, New York, 1943, 1970.

Hibbard, M.G., Dantzig, J.A., Trentham, D.R. and Goldman, V.E. "Phosphate Release and Force Generation in Skeletal Muscle Fibres." *Science* **228** (1985): 1317-9.

Ho, M.W. "How Rational Can Rational Morphology Be?" *Rivista di Biologia* **81** (1988a): 11-55.

Ho, M.W. "Re-animating Nature: The Integration of Science with Human Experience." *Beshara* **8** (1989a):16-25, (reprinted with minor revisions in *Leonardo* **24** (1991): 607-15.

Ho, M.W. "Coherent Excitations and the Physical Foundations of Life." In *Theoretical Biology. Epigenetic and Evolutionary Order from Complex Systems* (B. Goodwin and P. Saunders, Eds.), pp 162-76, Edinburgh University Press, Edinburgh, 1989b.

Ho, M.W. "A Quest for Total Understanding." *Saros Seminar on the Dilemma of Knowledge*, Transcript, Bristol Book Club, 1990a.

Ho, M.W. "The Role of Action in Evolution: Evolution by Process and the Ecological Approach to Perception.' In *Evolutionary Models in the Social Sciences* (T. Ingold, Ed.), special issue of *Cultural Dynamics* **4** (1991): 336-54.

Ho, M.W. "Natural Being and Coherent Society." In *Social and Natural Complexity* (E.L. Khalil and K.E. Boulding, Eds.), special issue of *J. Social and Biol. Structures*, 1992a.

Ho, M.W. "La Selezione dell'Amore." Scienza e Tecnologia, *L'Unita*, Dominica 26 Aprile, 1992b.

Ho, M.W. "Creativity and Reality in Art and Science." Public Lecture presented in Interalia Conference on Art and Science, Order, Chaos and Creativity, 29-30 Aug., Edinburgh, 1992c.

Ho, M.W. "Towards an Indigenous Western Science." In *Reassessing the Metaphysical Foundations of Science* (W. Harman, Ed.), Institute of Noetic Sciences, San Franscisco, 1993.

Ho, M.W. and Fox, S.W., Eds. *Evolutionary Processes and Metaphors*, Wiley, London, 1988.

Ho, M.W. and French, A., unpublished results, 1992.

Ho, M.W. and Lawrence, M. "Interference Colour Vital Imaging: A Novel Noninvasive Technique." *Microscopy and Analysis* (1993a) in press.

Ho, M.W. and Lawrence, M. "Interference Colour Vital Imaging of *Drosophila* Embryogenesis." in preparation, 1993b.

Ho, M.W., Lawrence, M. and Saunders, P.T. "Noninvasive Imaging by Interference Colours - Dynamically Ordered Regimes in Living Organisms." submitted for publication, 1993.

Ho, M.W., Matheson, A., Saunders, P.T., Goodwin, B.C. and Smallcomb, A. "Ether-induced Disturbances to Segmentation in *Drosophila melanogaster*." *Roux Arch. Devl. Biol.* **196** (1987): 511-21.

Ho, M.W. and Popp, F.A. "Biological Organization, Coherence and Light Emission from Living Organisms." In *Thinking About Biology* (W.D. Stein and F. Varela, Eds.), Addison-Wesley, New York, 1993.

Ho, M.W., Popp, F.-A., and Warnke, U., Eds. *Bioelectromagnetism and Biocommunication*, World Scientific, Singapore, 1994.

Ho, M.W., Ross, S., Bolton, H., Popp, F.A. and Xu, X. "Electrodynamic Activities and Their Role in the Organization of Body Pattern." *J. Scientific Exploration* **6** (1992b): 59-77.

190

190
190

Ho, M.W. and Saunders, P.T., Eds. *Beyond Neo-Darwinism : Introduction to the New Evolutionary Paradigm*, Academic Press, London, 1984.

Ho, M.W., Stone, T.A., Jerman, I., Bolton, J., Bolton, H., Goodwin, B.C., Saunders, P.T. and Robertson, F. "Brief Exposure to Weak Static Magnetic Fields During Early Embryogenesis Cause Cuticular Pattern Abnormalities in *Drosophila* Larvae." *Physics in Medicine and Biology* **37** (1992a): 1171-9.

Ho, M.W., Xu, X., Ross, S. and Saunders, P.T. "Light Emission and Rescattering in Synchronously Developing Populations of Early *Drosophila* Embryos - Evidence for Coherence of the Embryonic Field and Long Range Cooperativity." In *Advances in Biophotons Research* (F.A. Popp, K.H. Li and Q. Gu, Eds.), pp. 287-306, World Scientific, Singapore, 1992c.

Iwazumi, T. "High Speed Ultrasensitive Instrumentation for Myofibril Mechanics Measurements." *American Journal of Physiology* **252** (1987): 253-62.

Illich, E. *Tools for Conviviality*, Fontana, London, 1973.

John S. "Localization of Light." *Physics World* **44** (1991): 32-40.

Kell, D.B. and Hitchens, G.D. "Coherent Properties of the Membranous Systems of Electron Transport Phosphorylation." In *Coherent Excitations in Biological Systems* (H. Frohlich, and F. Kremer, Eds.), pp. 178-98, Springer-Verlag, Berlin, 1983.

Kyriacou, C.B. "The Molecular Ethology of the *Period* Gene." *Behavior Genetics* **20** (1990): 191-212.

Laing, N.G. and Lamb, A.H. "Muslce Fibre Types and Innervation of the Chick Embryo Limb Following Cervical Spinal Removal." *J. Embryol. Exp. Morphol.* **89** (1985): 209-22.

Li, K.-H. "Coherence in Physics and Biology." In *Recent Advances in Biophoton Research* (F.A. Popp, K.H. Li and Q. Gu, Eds.), pp. 113-55, World Scientific, Singapore, 1992.

Li, K.H., Popp, F.A., Nagl, W. and Klima, H. "Indications of Optical Coherence in Biological Systems and Its Possible Significance." In *Coherent Excitations in Biological Systems* (H. Fröhlich, and F. Kremer, Eds.), pp. 117-22, Springer-Verlag, Berlin, 1983.

Liburdy, R.P. and Tenforde, T.S. "Magnetic Field-induced Drug Permeability in Liposome Vesicles." *Radiation Res.* **108** (1986): 102-11.

Lomo, T., Westgarrd, R.H. and Kahl, H.A. "Contractile Properties of Muscle: Control by Pattern of Muscle Activity in the Rat." *Proc. Roy. Soc. Lond. (B)* **187** (1974): 99-103.

Lovelock, J.E. *Gaia: A New Look at Life on Earth*, Oxford University Press, Oxford, 1979.

Lovelock, J.E. *The Ages of Gaia*, Oxford University Press, Oxford, 1988.

MacDonald, R.D. (Trans./stage adaptation). *Goethe's Faust, Part I and II*, Oberon Books, Birmingham.

May, R.M. *Stability and Complexity in Model Ecosystems*, Princeton University Press, Princeton, New Jersey, 1973.

McClare, C.W.F. "Chemical Machines, Maxwell's Demon and Living Organisms." *J. Theor. Biol* **30** (1971): 1-34.

McClare, C.W.F. "A Molecular Energy Muscle Model." *J. Theor. Biol.* **35** (1972): 569-75.

McConkey, E.H. "Molecular Evolution, Intracellular Organization and the Quinary Structure of Proteins." *Proc. Nat. Acad. Sci. USA* **79** (1982), 3236-40.

McLauchlan K. "Are Environmental Magnetic Fields Dangerous?" *Physics World* **45** (1992): 41-5.

Meggs, W.J. "Enhanced Polymerization of Polar Macromolecules by an Applied Electric Field with Application to Mitosis." *J. Theor. Biol.* **145** (1990): 245-55.

Meister, M., Caplan, S.R. and Berg, H.C. "Dynamics of a Tightly Coupled Mechanism for Flagellar Rotation. Baterial Motility, Chemiosmotic Coupling, Protonmotive Force." *Biophys. J.* **55** (1989): 905-14.

Menzinger, M. and Dutt, A.K. "The Myth of the Well-Stirred CSTR in Chemical Instability Experiments: The Chlorite/Iodide Reaction." *Journal of Physical Chemistry* **94** (1990): 4510-4.

Mill, J. S. *Examination of Sir W. Hamilton's Philosophy*, 5th ed., 1878.

Morowitz, H.J. *Energy Flow in Biology*, Academic Press, New York, 1968.

Morowitz, H.J. *Foundations of Bioenergetics*, Academic Press, New York, 1978.

Musumeci, F., Godlevski, M., Popp, F.A. and Ho, M.W. "Time Behaviour of Delayed Luminescence in *Acetabularia acetabulum*." In *Advances in Biophoton Research* (F.A. Popp, K.H. Li and Q. Gu, Eds.), pp. 327-44, World Scientific, Singapore, 1992.

Needham, J. *Order and Life*, Yale University Press, New Haven, 1935.

Neurohr, R. unpublished observation, 1989.

Nicolis, G. & Prigogine, I. *Exploring Complexity*, R. Piper GmbH & Co., KG Verlag, Munich, 1989.

Nucitelli, R. "Ionic Currents in Morphogenesis." *Experientia* **44** (1988): 657-65.

Penrose, O. *Foundations of Statistical Mechanics, A Deductive Approach*, Pergamon Press, Oxford, 1970.

Penrose, O. "Entropy and Irreversibility." *Annals of the New York Academy of Sciences* **373** (1981): 211-9.

Penrose, R. *The Emperor's New Mind*, Oxford University Press, 1989.

Pethig, R. *Dielectric and Electronic Properties of Biological Materials*, Pergamon Press, New York, 1979.

Pethig, R."Protein-water Interactions Determined by Dielectric Methods". *Ann. Rev. Phys. Chem.* **43** (1992): 117-205.

Pohl, H.A. "Natural Oscillating Fields of Cells: Coherent Properties of the Membranous Systems of Electron Transport Phosphorylation." In *Coherent Excitations in Biological Systems* (H. Fröhlich, and F. Kremer, Eds.), pp. 199-210, Springer-Verlag, Berlin, 1983

Pollard, T.D. "The Myosin Crossbridge Problem." *Cell* (1987) **48**, 909-10.

Popp, F. A. "On the Coherence of Ultraweak Photoemission from Living Tssues. In *Disequilibrium and Self-Organization* (C.W. Kilmister, Ed.), pp 207-30, Reidel, Dordrecht, 1986.

Popp, F. A. and Li, K.H. "Hyperbolic Relaxation as a Sufficient Condition of a Fully Coherent Ergodic Field." In *Advances in Biophoton Research* (F.A. Popp, K.H. Li and Q. Gu, Eds.), pp. 47-58, World Scientific, Singapore.

Popp, F.A., Ruth, B., Bahr, W., Bohm, J., Grass, P., Grohlig, G., Rattemeyer, M., Schmidt, H,G. and Wulle, P. "Emission of Visible and Ultraviolet Radiation by Active Biological Systems." *Collective Phenomena* 3 (1981): 187-214.

Presman, A.S. *Electromagnetic Fields and Life,* Plenum Press, New York, 1970.

Prigogine, I. *Nonequilbrium Statistical Mechanics,* John Wiley & Sons, New York, 1962.

Prigogine, I. *Introduction to Thermodynamics of Irreversible Processes,* John Wiley & Sons, New York, 1967.

Rattemeyer, M. and Popp, F.A. "Evidence of Photon Emission from DNA in Living Systems." *Naturwissenschaften* 68 (1981): S572-73.

Ribary, U., Ioannides, A.A., Singh, K.D., Hasson, R., Bolton, J.P.R., Lado, F., Mogilner, A. and Llinas, R. "Magnetic Field Tomography (MFT) of Coherent Thalamocortical 40hz Oscillations in Humans." *Proceedings of the National Academy of Science* 88 (1991): 11037-9.

Rios, E. and Pizarro, G. "Voltage Sensor of Excitation-Contraction Coupling in Skeletal Muscle." *Physiological Reviews* 71 (1991): 849-908.

Sakurai, I. and Kawamura, Y. "Lateral Electrical Conduction Along a Phosphatidylcholine Monolayer." *Biochim. Biophys. Acta* 904 (1987): 405-9.

Saunders, P.T. and Ho, M.W. "On the Increase in Complexity in Evolution." *J. Theor. Biol.* 63 (1976): 375-84.

Schamhart, S. and van Wijk, R. "Photon Emission and Degree of Differentiation." In *Photon Emission from Biological Systems* (B. Jezowska-Trzebiatowski, B. Kochel, J. Slawinski, and W. Strek, Eds.), pp. 137-50,World Scientific, Singapore, 1986.

Schommers, W. "Space-time and Quantum Phenomena." In *Quantum Theory and Pictures of Reality* (W. Schommers, Ed.), pp.217-277, Springer-Verlag, Berlin, 1989.

Schrödinger, E. *What is Life?* Cambridge University Press, Cambridge, 1944.

Scott, A.C. "Solitons and Bioenergetics." In *Nonlinear Electrodynamics in Biological Systems*. (W.R. Adey and A.F. Lawrence, Eds.), pp. 133-42, Plenum Press, New York, 1984.

Sewell, G.L. "Non-equilibrium Statistical Mechanics: Dynamics of Macroscopic Observables." In *Large-Scale Molecular Systems: Quantum and Stochastic Aspects* (W. Gans, A. Blumen and A. Amann, Eds.), pp. 1-48, Plenum Press, 1991.

Slater, E.C. "Mechanism of Oxidative Phosphorylation." *Ann. Rev. Biochem.* 46 (1977): 1015-26.

Somogyi, B., Welch, G.R. and Damjanovich, S. "The Dyanmic Basis of Energy Transduction in Proteins." *Biochim. Biophys. Acta* 768 (1984): 81-112.

Stewart, I. "All Together Now..." *Nature* (News and Views) 350 (1991): 557.

Stryer, L. "The Molecules of Visual Excitation." *Scientific American* 257 (1987): 42-50.

Szent-Györgi, A. In *Light and Life* (W.D. McElroy and B. Glass, Eds.), Johns Hopkins Press, Baltimore, 1961.

Szent-Györgi, A. *Introduction to a Submolecular Biology*, Academic Press, New York, 1960.

Thornton, P.R. *The Physics of Electroluminescent Devices*, E. and F.N. Spon Lts., London, 1967.

Tien, H.T. "Membrane Photobiophysics and Photochemistry." *Prog. Surf. Science* 30 (1989): 1-199.

Vigny, P. and Duquesne, M. "On the Fluorescence Properties of Nucleotides and Polynucleotides at Room Temperature." In *Excited States of Biological Molecules* (J.B. Birks, Ed.), pp. 167-77, Wiley, London, 1976.

Welch, G.R., Somogyi, B. and Damjanovich, S. "The Role of Protein Fluctuations in Enzyme Action: A Review." *Progr. Biophys. Mol. Biol.* 39 (1982): 109-46.

Whitehead, A.N. *Science and the Modern World*, Penguin Books, Harmondsworth, 1925.

Williams, R.J.P. "On First Looking into Nature's Chemistry. Part I. The Role of Small Molecules and Ions: The Transport of the Elements. Part II. The Role of Large Molecules, especially Proteins." *Chem. Soc. Rev.* 9 (1980): 281-324; 325-64.

Winfree, A.T. *The Geometry of Biological Time*, Springer-Verlag, New York, 1980.

Winfree, A.T. and Strogatz, S.H. "Singular Filaments Organize Chemical Waves in Three Dimensions. I. Geometrically Simple Waves." *Physica* 8 D (1983): 35-49.

Zajonc, A.G. "New Consciousness, New Thinking and the New Physics." Paper presented at Conference on *Re-assessing the Metaphysical Foundations of Modern Science*, Institute of Noetic Sciences, Asilomar, California, Nov. 20-24, 1991. To appear in proceedings.

SUBJECT INDEX

AUTHOR INDEX